GUT RENOVATION

Also by Roshini Raj, MD

What the Yuck? The Freaky and Fabulous Truth About Your Body

GUT RENOVATION

UNLOCK THE AGE-DEFYING POWER OF THE MICROBIOME TO REMODEL YOUR HEALTH FROM THE INSIDE OUT

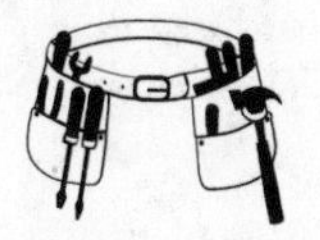

ROSHINI RAJ, MD

with Sheila Buff

WILLIAM MORROW

An Imprint of HarperCollins*Publishers*

This book contains advice and information relating to health care. It should be used to supplement rather than replace the advice of your doctor or another trained health professional. If you know or suspect you have a health problem, it is recommended that you seek your physician's advice before embarking on any medical program or treatment. All efforts have been made to assure the accuracy of the information contained in this book as of the date of publication. This publisher and the author disclaim liability for any medical outcomes that may occur as a result of applying the methods suggested in this book.

HarperCollins books may be purchased for educational, business, or sales promotional use. For information, please email the Special Markets Department at SPsales@harpercollins.com.

A hardcover edition of this book was published in 2022 by William Morrow, an imprint of HarperCollins Publishers.

FIRST WILLIAM MORROW PAPERBACK EDITION PUBLISHED 2023.

Library of Congress Cataloging-in-Publication Data has been applied for.

ISBN 978-0-06-314421-7

23 24 25 26 27 LBC 5 4 3 2 1

To my Dad—you are with me every step that I take.

And to my Mom, Kiren, Dilan, and Manish. Your love gave me the energy to bring this book to life.

CONTENTS

Introduction 1

Chapter 1 Meet Your General Contractor: Your Microbiome 5

Chapter 2 The Architect: Your Brain 21

Chapter 3 The Kitchen: Eating Your Way to a Younger Body and Mind 35

Chapter 4 The Bathroom: Eliminating Issues 61

Chapter 5 The Powder Room: Beauty Isn't Just Skin Deep 85

Chapter 6 The Home Gym: Work Out to Turn Back the Clock 101

Chapter 7 The Zen Corner 113

Chapter 8 The Bedroom: Sleeping Your Way to Better Health 129

Chapter 9 The Nursery: Healthy Guts, Healthy Kids 145

Chapter 10 The Laundry Room: Detoxing Your Home 159

Chapter 11 The Living Room: The Gut Reno Program 173

Gut Reno Workout Week 187

Gut Reno Recipes 197

Acknowledgments 247

Notes 249

Index 265

GUT RENOVATION

INTRODUCTION

A question I get asked all the time is "Why did you decide to become a gastroenterologist?" Strange as it may sound, I love looking inside my patients' bodies, seeing problems, and fixing them on the spot. This is especially meaningful when I'm dealing with a serious issue, like a cancerous colon polyp or a bleeding stomach ulcer. But even in situations that aren't life-threatening, I find it fascinating to explore the inner workings of the human body and see how our amazing natural machinery functions.

In my practice, I treat both men and women, but because many women seek out a female gastroenterologist, the majority of my patients happen to be women. As a woman who has gone through puberty, pregnancy, and childbirth, and who hears menopause knocking on the door, I can personally relate to many of the symptoms my patients experience. I too have eaten something iffy and then spent the evening running to the bathroom. Like most women, I know what it feels like to be bloated, nauseated, constipated, and crampy. And thanks to giving birth to my two kids, I understand what it's like to have painful hemorrhoids. (To be clear, I'm not calling my children hemorrhoids—at least not to their faces.)

As a gastroenterologist and internist at NYU Langone Medical Center, I examine people literally from head to toe. I can see how they're aging all over, including from the inside when I perform endoscopies or colonoscopies. It's not uncommon for me to see a fifty-year-old patient with the medical issues of a seventy-year-old, or vice versa. And appearances can vary greatly. I often have to do a double take as I reconcile the age of the person I read in their chart and the age they appear when I see them standing before me. I'm always curious to find the reasons and behaviors that may be accelerating or slowing their aging process. I want to help those who are aging too fast, and learn from those who maintain youthful vigor and good health as they age. And let's face it, I have another motive. As a woman who just celebrated her fiftieth birthday you better believe I'm *very* interested in gleaning the secrets to aging well, both internally and externally.

Every day, I see firsthand the struggles people face as the result of poor diet and lifestyle choices. Many people think the way they age or how long they'll live is out of their hands, but that's just not true. Genetic factors are responsible for only 25 to 30 percent of your expected longevity, which means you have a lot more control over how well or how poorly you age than you may think. This may sound intimidating, but it is good news! The key is managing your gut health.

Your gut health matters in ways you may not expect. The changes I recommend in *Gut Renovation* can lead to real improvements in digestive health, chronic health problems, cancer prevention, and mental outlook. A balanced microbiome is also the foundation for enhancing your immunity and protecting against the age-related chronic diseases that could lead to premature aging, years of poor health, and a shortened life span. In other words, with the right approach to your digestive health, you can age in *slo-mo*!

As much as I've always loved being a gastroenterologist, the best part of the job has emerged in the last few years. An explosion of research has shown us the staggering power of the gut microbiome—that teeming community of bacteria, viruses, and yeasts residing in your gut and extending its influence throughout your body.

But how exactly does your microbiome function within your body? How can changing your microbiome help slow the aging process? What role does it play in your immunity? More importantly, how can you optimize your microbiome to promote your healthiest, most youthful self? That's where this book comes in—to help you take this amazing science, apply it to your daily life, and achieve its maximum benefits to age well, look great, and live better.

In *Gut Renovation*, I present my game-changing program for improving your microbiome and boosting your digestion, your immunity, your mood, and your vitality. I combine my practical clinical experience with the latest microbiome research to provide clear, easy-to-grasp advice that will allow you to understand—and make!—the necessary changes to reset your biological clock. The key? Giving your microbiome a room-by-room makeover.

I know, it sounds wild—but hear me out. I often tell my patients to think of their bodies as a house. To stay safe and comfortable, a house needs constant maintenance and basic repairs. But to turn your house into the home of your dreams, it needs a full renovation. You'll tear out old, outdated fixtures, restore some rooms, expand others, and redecorate. The process might be a bit messy and take some investment of time and energy, but in the end, when your house has become a pleasurable place to inhabit for a long time, it's worth it.

In *Gut Renovation* I distill everything I've learned during my many years as a gastroenterologist and condense the strategies I've developed for my patients to look and feel their best. I personally practice all of these approaches in my own life and have seen the beneficial changes that occur. With the right tool kit, we all can feel great, look forward to a dynamic future, and literally get better with age. It's never too late to correct your course and reset the biological aging process—and the sooner you do it, the better. That's why starting with this *literal* Gut Renovation now positions you for the best, brightest future.

The plan I lay out in *Gut Renovation* will help you take good care of your inner ecosystem, protect your cells from damage, and kick-start the cellular repair process. I explain the pivotal role your microbiome plays in every aspect of your health and how our modern diet and lifestyle is putting our gut balance in jeopardy.

As a physician, I must tell you that while most of the strategies you read here will be beneficial for everybody, it's best to consult your doctor before making any major changes in your diet, exercise routines, or lifestyle.

In many ways, the timing of your renovation couldn't be better. One of the main effects of the COVID-19 global pandemic is that we all had to reckon with our health and mortality in new ways. It's made us aware of changes that needed to be made in our diets, our exercise routines, and stress management tactics. Now, more than ever, it is imperative that we make our health a priority. This is the time to build our resilience to disease and aging. We also need to deal with "pandemic pounds," skyrocketing mental health issues, long-haul post-COVID symptoms, and the other unhealthy sequelae of time spent in quarantine conditions, all issues that doctors like myself are now seeing in our offices. Now really is the perfect time to reinvent, reinvigorate, and reset your aging clock.

But you won't be alone on this journey! I'm here to guide you room by room through the remodel of your gut and your health. We'll put you on the path toward a more vibrant future by changing the way you age on a cellular and microbiological level. The end result of your Gut Renovation will make you healthier and more youthful, and set you up to live a longer, happier life. That's a pretty ambitious scope of work, I know, but I'm confident we can get it done. Let's get to it!

CHAPTER 1

MEET YOUR GENERAL CONTRACTOR: YOUR MICROBIOME

Before we dive into your room-by-room Gut Renovation, I want to introduce you to your general contractor, the head honcho who's calling the shots in this overhaul: your microbiome. The microbiome is an overall term for the many trillions (yes, trillions) of microbes (bacteria, viruses, and fungi) that live in and on your body. You can't see them, but your microbiome is living in and on every part of you—from your skin to your genitals to your colon. Even your ears and eyes have their own microbiome!

Much as the general contractor leads every aspect of a home renovation, the microbiome influences almost every part of your health. And as with a general contractor, a lot of this work happens behind the scenes! All of those microorganisms in your microbiome are an integral part of you. And there are a *lot*. You have somewhere around twenty thousand to twenty-five thousand human genes; your microbiome has as many as eight million genes. In your gut alone, your microbiome makes up about four pounds of biomass. Compare that to your brain, which weighs only about three pounds. For as long as humans have existed, so have the bacteria on us and in us. You're not just a person, you're a superorganism! Your assortment of microbes is different from anyone else's, even from a twin—your microbiome is as unique as a fingerprint.

How do we know this? In 2007, the National Institutes of Health launched a research initiative called the Human Microbiome Project, which aimed to identify and characterize the composition of the human gut microbiome. Thanks to those brave researchers sifting through a ton of stool, the discoveries from the project not only led to a much greater understanding of what's brewing inside our guts, but also jump-started game-changing research into how these bugs function in our body.

So what are all those bacteria doing?

A LOT. Basically, they're keeping you alive and healthy. In your gut (your small intestine and colon), they're performing vital functions that affect your entire body. Your gut bacteria help you digest your food, extract the nutrients you need to survive, and produce some of the vitamins your body depends upon. Pretty useful houseguests, aren't they? They also bolster your immune function, make anti-inflammatory compounds, produce neurochemicals that affect your mood and cognition, and support your health in myriad other ways—as you will be learning throughout this book.

More than two thousand years ago, Hippocrates said, "All disease begins in the gut," and modern science seems to be proving him right. Research shows that your beneficial bacteria can influence everything from your body weight to your risk of developing obesity-related diseases (like type 2 diabetes), chronic inflammatory disorders (such as inflammatory bowel disease), heart disease, mental health problems (including depression and anxiety), and age-related musculoskeletal conditions (like osteoporosis and sarcopenia).

What has also become clear is that the connection works in both directions. Your gut microbiome not only affects other parts of your body and your well-being, but your microbiome is also *affected by* your lifestyle and your overall health.

Why Diversity Matters

We often associate bacteria with infection and illness. I certainly do—in my office we spend a lot of time and effort on washing our hands, cleaning surfaces, and sterilizing equipment to get rid of bacteria. In any setting where harmful bacteria can be passed on, that makes sense.

Your gut is different. The wide variety of bacteria in the human microbiome consists of mostly beneficial bacteria, not harmful ones. (Otherwise, you'd be in trouble.) Your microbiome naturally goes through ups and downs, blooms and die-offs, in bacterial species and in numbers. A healthy gut has about 85 percent beneficial species, which means that a substantial number of harmful bacteria are always hanging around looking for a chance to multiply and throw the balance off. In a healthy gut, if that happens you might have a day or two of feeling a little off, but usually the balance fixes itself without you even noticing.

You can never fully get rid of the bad—or what I call unfriendly!—bacteria, but with the right care and habits you can maximize the effectiveness of good bacteria. That way, the bad guys will be crowded out by the good guys so they can't reproduce into large enough numbers to be harmful.

Sometimes the unruly houseguests—the unfriendly bacteria—do get the upper hand and trash the place. That might be because you unknowingly ate food contaminated with bad bacteria like salmonella or *E. coli* and got food poisoning. Or maybe you caught a virus—what we doctors very scientifically call a stomach bug. After a few days of throwing up and/or diarrhea, your immune system clears out the bad guys, the good bacteria take over again, and your digestion gets back to normal.

At least, that's what usually happens. Sometimes, however, the bad bacteria linger, and your digestion takes a long time to return to normal. Or the bad bacteria don't cause an actual illness but still crowd the good bacteria, reducing their numbers. Your microbiome is incredibly dynamic, changing quickly in response to your diet and environment. And so consistent bad habits mean that the balance of your microbiome in your gut is regularly set off course. As the balance and diversity of bacteria deteriorates, so does your health, because you now have a problem called dysbiosis.

Understanding Dysbiosis

Modern life doesn't make things easy for your gut microbiome. We're constantly doing things to it that the bacteria don't like. Top of the list is our diet. Many of us eat the aptly named Standard American Diet, or SAD. It's mostly high-calorie, low-nutrition, ultraprocessed foods filled with sugar, salt, bad fats, preservatives, and food additives. This sort of junk food now makes up more than half of the average American's diet. On top of a bad diet, we attack our gut bacteria with alcohol, antibiotics and other drugs, environmental toxins, lack of sleep, and plenty of stress. Those lifestyle choices can all throw your microbiome into an imbalance that can't self-correct. You feel your digestive system just isn't working right, but because your symptoms are vague and variable, you ignore them or maybe pop some over-the-counter pills.

Dysbiosis is basically what happens when there's a major reduction in the diversity of your microbiome and/or an overgrowth of unfriendly bacteria. The symptoms can vary a lot from person to person, and from day to day, even hour to hour, in the same person. The most common digestive symptoms include upset stomach/nausea, constipation, diarrhea, and gas and bloating. You may also have fatigue and brain effects such as brain fog, inability to concentrate, anxiety, and depression.

Not everyone will have every symptom, of course, and the symptoms can range a lot in how severe they are and how often you have them. But even without any digestive symptoms, dysbiosis can wreak havoc on your body in many ways. It has been implicated in everything from skin disorders to diabetes.

Throughout this book we'll talk about how your microbiome and gut health interact with all the aspects of your life, and the strategies you can employ to prevent or fix dysbiosis. Two important tactics are improving your diet and using probiotics—good bacteria—to restore a better balance in your microbiome. But there are many more ways to optimize your gut, as you will soon learn!

Gut Barrier Function

When food reaches your small intestine in its digestive journey, it's basically a soup of partially digested food, as well as anything toxic you may have accidentally eaten. (I know, gross!) While you want the nutrients from your food to be absorbed into your bloodstream, you don't want to absorb anything else. Lining your small intestine is a single layer of cells closely laid next to each other, sort of like subway tiles, with very little space between them. The spaces between the cells are called tight junctions. The junctions can open up just enough to allow digested food particles, water, and micronutrients to pass through into the bloodstream, while blocking larger particles and the rest of the intestinal contents. In this way, the lining of your intestine forms an important barrier. It keeps the toxic or foreign particles in your intestine from sneaking into your bloodstream, where they can trigger inflammation. The good bacteria in your gut help maintain this barrier, both by secreting protective mucus to layer onto the gut wall and by producing compounds that keep the junctions tight.

But what if that barrier gets compromised? When the tight junctions of your small intestine open up too widely or stay open too long, or if the delicate walls of your small intestine develop tiny holes and cracks, you have intestinal permeability—aka a leaky gut.

When the gut becomes permeable for whatever reason, larger food particles, bacteria, and other intestinal contents leak out into your bloodstream. Your immune system sees the escaped contents and responds to them as if they were dangerous invaders, which in a sense they are. The immune response triggers inflammation, which in turn can lead to many of the same symptoms as dysbiosis. You can get bloating, gas, nausea, and cramps, but now with the possible addition of food sensitivities and aching joints. Long term, it's possible that the chronic inflammation caused by a leaky gut can lead to autoimmune illnesses such as rheumatoid arthritis or other chronic diseases like diabetes or even heart disease. It can also produce food allergies that won't go away.

What causes these breaches in the integrity of the intestinal lining? One cause is dysbiosis that goes on for a long time, but there are many others. The same low-nutrient, low-fiber diet that causes dysbiosis can cause leaky gut by constantly bombarding your gut with damaging substances, including the artificial sweeteners, preservatives, food additives, food colorings, emulsifiers, and residue of agricultural chemicals that abound in processed and packaged foods. Alcohol can damage the gut, as can the many environmental toxins that we're exposed to every day: air pollution, cleaning supplies, cosmetics, personal care products, fire retardants, fabric softeners, and everything else. Having a bad bout of food poisoning or a stomach flu can also increase intestinal permeability.

Radiation therapy for cancer and a range of powerful drugs for cancer and other serious conditions can also trigger leaky gut. If you have an underlying condition, like celiac disease or Crohn's disease, you've got a good chance of developing a leaky gut, because the inflammation from these diseases can directly damage your intestinal lining. And then there's that all-purpose enemy of digestion, stress.

So while a leaky gut can be caused by many factors, you have control over so many of them! You're the general contractor of your microbiome, after all. To start, changes in gut flora need to occur to correct dysbiosis. Restoring the numbers of good bacteria and the diversity of species leads to tighter junctions in the gut lining as well as more protective mucus—both of which strengthen and reinforce this important barrier.

Your Home Surveillance System: Understanding Immunity

From start to finish, your digestive tract is constantly in contact with bacteria. Keeping those bacteria inside the long digestive tube is important to protect the rest of your body from infection. Some are inevitably going to escape, however. Your body is ready for them: at least 70 percent of all the infection-fighting immune cells in your body are found in the gut. You can think of them as your body's alarm system.

Having an alarm system is a good thing, right? After all, you want your gut immune system to go after the bad microbes. But as anyone with a home alarm system knows, it's easy to set off a false alarm. You don't want your immune system going off by mistake, because that's what triggers autoimmune diseases, where the body attacks itself. Ideally, you want your immune system to be balanced between tolerating a few bad microbes and reacting swiftly when the level of bad microbes reaches the danger point.

We call this immune tolerance. The best way to maintain it is by having a diverse range of gut bacteria. Diversity helps the cells of your immune system distinguish between dangerous microbes that need to be attacked and those that don't—and to distinguish between invaders and your own cells.

When your immune system does need to respond, a complex cascade of steps gets triggered. Imagine you're making dinner and you slice your finger while chopping the onions. That's sort of like an alarm system when one of the safety protocols is breached. Your body's alarm is quick to respond; bacteria from the environment immediately enter the gash and your immune system kicks in to throw them out. The area around the cut swells, gets red, feels hot, and hurts, all of which are signs of acute inflammation. Basically, the first immune cells to rush to the scene send out chemical signals that tell the blood vessels around the cut to get leaky—the cells that line the blood vessels open up a little to let more immune cells, platelets (for clotting), and fluid from your blood into the area. That makes the area around the cut swell even more.

Just as an alarm system would signal for backup, so does your body. The immune cells that rush in start sending out more chemical messengers called cytokines. The cytokine messages help control the inflammatory response. They tell more immune cells to come to the cut to kill off invaders.

If the cut is a small one, your immune system can easily kill off any invading bacteria. Your finger may be a little red, swollen, and sore for a few days until it heals up. But if the cut's a bad one, or if you just get unlucky and particularly dangerous bacteria get in, your finger could develop an infection. Now your immune system has to work harder to get rid of the invaders. The cytokines tell more immune cells to join in, and also make you run a fever and feel tired so you'll slow down and have more energy for fighting the infection.

Acute inflammation may make you feel pretty lousy for a few days, but when the worst is over, the inflammatory response dies down.

The tricky thing about acute inflammation in the gut is that it isn't visible the same way a cut on the finger is, but your body has a lot of the same recovery mechanisms. Your gut gets swollen and painful with cramps and bloating. It doesn't work well, so you may get diarrhea and may even pass some blood. You feel fatigued and achy, lose your appetite, and run a fever.

Your body is *constantly* on the lookout for things that can harm you. Your immune system goes after bacteria and viruses that can make you sick, but it also attacks anything else it sees as an invader, like particles of undigested food that can escape through a leaky gut.

What if the inflammation isn't from an infection but from dysbiosis, a leaky gut, obesity, an allergy, an autoimmune disease, or some other cause that puts your immune system into a state of constant, but low-level, alert?

Things escalate. Those cytokines, which are only supposed to be released when they're truly needed, are now being sent out into your bloodstream constantly. That's chronic inflammation—the kind that lingers on, causing ongoing, low-grade symptoms. Chronic inflammation in the gut gives you unpleasant symptoms, makes you feel tired all the time, gives you rashes and vague aches and pains in your joints and muscles, and can make you feel irritable and foggy, even depressed. Left untreated, it can start damaging healthy tissues, like your arteries, joints, and brain.

Inflammaging and Your Gut Microbiome

When most people think about aging, they usually think about gray hair and wrinkles, or maybe they also see getting older as a time when they'll have less strength, vitality, and mental sharpness. And trust me, as I've recently celebrated my fiftieth birthday, these signs of aging are top of mind for me too! But being a doctor and seeing how differently people can age within my pool of patients has given me a broader perspective on aging. When I compare an eighty-year-old patient who literally hops onto the examining table like she has springs in her shoes to a fifty-two-year-old patient who listlessly shuffles into the office using a cane and wearing an air of exhaustion, I am reminded that aging doesn't treat us all equally. I see aging as not only the normal wearing down of your body, but also as the increased susceptibility to organ damage and disease. In other words, what makes us "old" is usually the conditions we accrue as we age—things like cancer, heart disease, type 2 diabetes, and arthritis, to name a few.

And what makes us more susceptible to these conditions as our body ages? Turns out it's a phenomenon called *inflammaging*. This refers to the low level of inflammation that increases in our bodies as we get older. When we look at the level of inflammatory markers (chemicals in the blood that rise during inflammation) in different age groups, we see a two to four times increase in the elderly. But not everyone develops the same degree of inflammaging as they age, and we see that healthier older people tend to have less inflammation and suffer from fewer diseases.

When chronic, low-grade inflammation goes on for a long time, it can negatively affect everything from brain circuitry to hormones to organ function to tumor formation. You become more likely to develop an age-related disease like Alzheimer's, coronary artery disease, and severe osteoarthritis—and that disease is more likely to hit you at a younger age than it would someone without chronic inflammation. What's interesting is that, in effect, *chronic inflammation accelerates the aging process*. And this happens on many levels.

In fact, you can predict someone's age with startling accuracy (within four years) just by studying the amounts and types of bacteria residing within them—and inflammaging is a big part of that. Inflammaging can also accelerate outward signs of aging like frailty, loss of mobility, muscle loss, and skin wrinkles. So both in terms of how old you look and feel, inflammaging plays a crucial role.

Much of the chronic inflammation that causes inflammaging originates in the gut. As you get older, the makeup of your gut bacteria naturally changes and tends to become less diverse—and less diversity in the gut microbiome, along with other issues, means more inflammation throughout the body. Recent research into the microbiome shows that people of different ages tend to contain a different community of bacterial species in their guts. The uniqueness of your microbiome tends to kick in more after age forty, and it varies from person to person. When scientists recently compared the gut bacteria of people of all ages ranging eighteen to ninety-eight, they found that certain specific bacterial profiles were associated with healthier aging. In fact, they found a decreased four-year survival for people with higher amounts of a bacteria called *Bacteroides* and lower overall uniqueness of their microbiome.[1]

Research studying centenarians and supercentenarians (104 years and older, bless them!) shows a similar effect. They tend to have more diverse microbiomes compared to younger and less healthy individuals. This concept of a microbiological clock is not only fascinating, but the implications are clear: if we can lower inflammation levels, we can manipulate our microbiome composition to a "younger" profile, which means, yes, we might actually slow down or even reverse some of the aging process. A huge part of this—as you'll learn in a couple of chapters!—comes down to diet.

COVID-19 and Your Gut

In the process of fighting COVID, your immune system can get so stimulated that it causes a runaway reaction. If your body is already inflamed by dysbiosis, inflammaging, or chronic disease such as type 2 diabetes, your immune system may already be primed to overreact.

Why can't your body put the brakes on and slow down your immune response? The answer may lie in the types of bacteria in your gut. In fact, a recent study on the gut microbiomes of people with and without severe illness from COVID suggests that the ones with the most diversity in their gut bacteria were the most likely to have mild disease, because they had plenty of the good bacteria that we know help regulate your immune system. The ones with the least diversity tended to get the sickest because their immune system was already overstimulated. The sickest people tended to have higher-than-usual levels of some bacteria species that are associated with inflammation, and lower levels of some bacteria species that are associated with a normal immune response. In addition, the people who got the sickest continued to have lower than usual diversity in their gut microbiome for months after they recovered.[2]

We know that COVID infects the digestive tract as well as the lungs, and the virus can be isolated from stool samples long after it disappears from nasal swab tests. This is one reason why the Chinese government chose an anal swab as their test of choice for foreign travelers—much to those visitors' dismay! We're also seeing many gastrointestinal effects from COVID, both acutely and long-term. One study showed that more than 30 percent of patients with COVID experienced GI symptoms like nausea, diarrhea, and loss of appetite during their initial illness.[3]

I have seen several patients in my practice who, several months after recovering from COVID, still have persistent bowel issues like bloating, abdominal pain, and diarrhea. In fact, practicing in NYC where we experienced rampant COVID in the early days of the pandemic, one of the first questions I now ask when I see a patient with almost any digestive symptom is: did you have COVID? It's a frustrating situation, and we don't yet know all the answers in terms of treatment, but I have seen good success using strategies to restore a balanced microbiome (like those in this book!) to help these patients feel better.

Unfortunately, we still have a lot to learn about how COVID affects the body. People who have recovered from COVID have depleted diversity in their gut bacteria for a long time afterward. Does this help explain why some people have "long COVID," with symptoms that linger on? Would restoring their bacteria to a better balance help them recover to a pre-COVID microbiome state? Would taking a probiotic help prevent contracting severe COVID? This is an active area of research that should provide answers soon. One British study of more than three hundred thousand people found that for women, regularly taking probiotics decreased their risk of testing positive for COVID.[4]

COVID in some form and its aftereffects are likely to be with us for a long time, even with the vaccines. In addition to the precautions we're all taking, like washing our hands, we can now see that having a healthy gut microbiome is going to be more important than ever. When your gut is healthy, you're less likely to get sick in general and you're more resilient when you do—you don't get as seriously ill, and you bounce back quickly.

Probiotics, Prebiotics, and More!

Gut health is having its moment, so you're probably hearing people talk about probiotics and prebiotics. The terms are often used very loosely, which leads to a lot of confusing misinformation. Let's define exactly what they mean.

PROBIOTICS

Probiotics are live bacteria that confer a health benefit to the host (*pro* = for, *biotic* = life). As the name indicates, these bacteria are *good* for you—and pretty much everyone can benefit from them. We can directly add them to our gut by eating probiotic-containing foods or by taking supplements. Probiotic supplements usually contain several different bacteria species, usually strains of the groups called *Lactobacillus* and *Bifidobacterium*—the predominant beneficial bacteria in the human gut. Probiotic supplements may also contain other bacteria; some also contain a yeast called *Saccharomyces boulardii*. Swallowing the supplement will send the bacteria to your colon, where they will take hold, reproduce, and help create a bigger population of the good guys.

The product may include some specific strains within a bacterial species, with confusing names like *L. plantarum* LP-115. The idea is that some specific bacteria and bacterial strains work better than others at colonizing your gut. Some manufacturers even claim that specific strains can help with specific issues, like depression. Is this true? The research is intriguing, and I fully believe that in the future we will see probiotic supplements tailor-made for specific diseases—but apart from a few pioneers, we still need more science to back up most of these disease-specific claims. There are also companies that create personalized probiotics based on a stool-based microbiome analysis. They claim to be able to determine which bacterial species you're missing and replenish those strains with a custom-made supplement. Again, I think it's a little early to believe the promise here, but I do think in the future we will see very valid offerings in this space.

When choosing a probiotic supplement, look for one that has at least one billion CFUs (colony forming units) per capsule—many products contain ten or even fifty billion CFUs, but more isn't always better. The CFU count, also sometimes called the live cultures, is the number of live probiotic bacteria in the formula. (They're dormant until they reach your colon.) Watch out for products that list the CFU at time of manufacture—some will naturally die off before you swallow them, so you want to know how many live bacteria the product will have when it reaches its expiration date. The label should state the product was made using good manufacturing processes (GMP). The "coolest" probiotics? Many patients ask me if probiotics need to be refrigerated, and the answer is, not necessarily. Some of the best probiotic companies use freeze-dried technology that allows their probiotics to remain alive for a long time at room temperature.

Probiotics are helpful when you need to take an antibiotic, because they help restore the beneficial bacteria that get killed off along with the harmful ones.[5] They're also very helpful for a nasty, hard-to-treat bacterial infection of the colon called *Clostridioides difficile*, or *C. diff*, that you can get after taking antibiotics for a long time. In my practice, I often recommend probiotics for my patients with irritable bowel syndrome and other bowel issues as a way to manage symptoms—they even help some patients put their disease into remission.[6] But this is just the beginning. Several studies have shown that probiotics can improve cholesterol profiles, and recently the FDA sanctioned a probiotic supplement to help control blood sugar levels in people with diabetes.[7] We're seeing and will continue to see the expansion of the use of probiotics well beyond digestive diseases.

Probiotic supplements can be helpful, but in order to get a really good variety of bacteria, you can't rely on just one supplement. That's why I always recommend also eating fermented foods that contain live bacteria, because some of the bacteria that drive the fermentation process will get carried into your colon. Good choices are live-culture yogurt and kefir, miso, tempeh, sauerkraut, kimchi, and lacto-fermented pickles (the kind made with salt, not vinegar).

PREBIOTICS

Whenever you take probiotics, I recommend also taking supplemental prebiotics. Prebiotics are substances that fuel the growth of your beneficial bacteria (probiotics). If probiotics are the blooms, then prebiotics are the fertilizer that helps them thrive. Prebiotic supplements contain soluble fiber, usually in the form of fructo-oligosaccharides (FOS) and inulin, complex sugars that pass undigested through the small intestine and arrive in the colon like a birthday cake for your bacteria.[8] As with probiotics, look for a statement about GMP on the label.

In addition to prebiotic supplements, you should eat foods that are high in oligosaccharides and inulin. They're found naturally in a lot of plant foods, including almonds, asparagus, avocado, barley, berries, cabbage, cherries, chia seeds, chickpeas, coconut, garlic, greens, Jerusalem artichokes, lentils, onions, peaches, pistachios, and walnuts.

SYNBIOTICS

When your gut bacteria are out of balance enough to be causing symptoms, we can use supplements containing probiotics and prebiotics to help restore beneficial bacteria and improve the diversity of species. To make life a little simpler, many manufacturers now make synbiotics—supplements that contain both prebiotics and probiotics in one capsule. Why? Because the effect of combining a probiotic with a prebiotic is synergistic. The two are more powerful together than each is alone.

POSTBIOTICS

Just when you thought you had a handle on the terminology, there's a new kid on the block: postbiotics. When the good bacteria in your gut digest fiber, they release bioactive compounds—metabolites—that have a beneficial effect on you. Postbiotic supplements are a step beyond probiotics because they contain just the beneficial metabolites, not the live bacteria. Postbiotics benefit you long after the bacteria that produced them have died off.

One caveat to all this is that, unfortunately, the supplement industry isn't as rigorously regulated as it should be. Product labels don't always accurately reflect what's actually in the bottle. Recommendations from your doctor or going with bigger, more reputable manufacturers will help ensure you actually get what you're paying for.

So as you can see, your microbiome really is the key to your overall renovation. As your general contractor, your microbiome affects every aspect of your health from your immunity, the level of inflammation in your body, to how well you age and how likely you are to develop chronic disease. In the following chapters we'll dive into how to keep your microbiome healthy and happy so that it can take the best care of you.

CHAPTER 2

THE ARCHITECT: YOUR BRAIN

Now that you know all about your general contractor—your microbiome—you need to meet the other key player coordinating your renovation: the architect, aka your brain. Do you ever experience butterflies in your stomach during a stressful time? Or maybe you have a gut feeling that something isn't right? Or feel a gut-wrenching emotion? Those sensations are a great example of how your gut and your brain are tightly linked—so much so that they actually determine much of your daily functioning.

Amazingly, the microscopic bacteria in your gut can have a powerful impact on your brain: your thinking, your mood, your mental health, and your risk of cognitive decline are all linked to the microbiome. And the converse is also true. Your brain "talks" to your gut, sending signals that govern many of the steps in the digestive process. And just as communication between an architect and a contractor can break down, a Gut Renovation can get stalled. If the interplay between the gut and the brain is out of whack, your health suffers. But first let's learn about how these guys can work together in harmony.

The Gut–Brain Axis

Most of what your body does to keep you alive is completely out of your conscious control. This is thanks to your autonomic nervous system, the part of your nervous system based in your brain stem and spinal cord that handles all your basic body functions, like your breathing, your heartbeat—and your digestion.

What happens after you swallow a bite of food, for example, is handled automatically by your enteric nervous system (ENS). This vast, complex network of at least two hundred million neurons lines the entire length of your digestive tract, from your mouth all the way to the anus. Because it's so huge and impactful, the ENS is sometimes called your second brain. Behind the scenes, without you needing to do anything, the ENS and your brain are in constant communication. Most of what they're talking about is how to coordinate your digestive system with other systems in your body, such as your immune system, to keep everything moving along without any unnecessary hiccups like inflammation—or actual hiccups, for that matter!

Another important link between your gut brain and your real brain is your vagus nerve. Also known as the tenth cranial nerve, the vagus nerve originates in your skull at the back of the brain and takes a long, meandering path through your body. As it wanders through your upper body, branches of the vagus touch your throat, your heart, your lungs and diaphragm, and most parts of your digestive system, including the stomach, small intestine, and part of the colon, as well as your liver, gallbladder, and pancreas.

Gut–Brain Crosstalk

We're still learning exactly how the gut and the brain talk to each other—and what the conversation is all about. One pathway for exchanging information is hormones and neurotransmitters; another is your immune system; yet another is metabolites—molecules made by the bacteria in your microbiome. And then there's that connecting highway, the vagus nerve.

Let's start with hormones and neurotransmitters. Specialized cells in your gut produce some twenty different hormones that send chemical messages from the gut to the rest of the body, including your brain. In your stomach and small intestine, for example, you produce a hormone called ghrelin that plays a big role in controlling how much you eat. Ghrelin is often called the hunger hormone because it stimulates your appetite and promotes fat storage. It works by circulating through your bloodstream and acting on an area of your brain called the hypothalamus that's important for appetite control.

The opposite of ghrelin is leptin, a hormone that's mostly produced by your fat cells (small amounts are also produced in your gut). Leptin is the satiety or "I feel full" hormone, the one that tells your hypothalamus you've had enough to eat and it's time to turn off the hunger pangs. Your ghrelin and leptin levels are always in a sort of dance with each other; what makes one go up lowers the other, and vice versa. As you can imagine, alterations in these hormones can play a big role in weight gain and food addiction.

But how do the hormone-producing cells in your gut know when to send a message to the rest of the body? Your bacteria tell them. The metabolites made by your gut bacteria provide the hormone-producing cells with information about what's going on in your gut: what you've been eating, how well your gut lining is holding up, the composition of the bacterial population itself, and more. In response, your gut releases the appropriate hormones in the appropriate quantities. If this signaling mechanism by the bacteria is thrown off, perhaps by dysbiosis, then inappropriate amounts of hormones can be released. In the case of ghrelin, for example, that can lead to overeating and obesity.

Going in the opposite direction, your brain stimulates your pituitary gland and the hypothalamus to work with your adrenals (little glands that sit on top of your kidneys) to release the stress hormone cortisol when you feel threatened. That cortisol surge diverts blood away from your gut in preparation for the fight-or-flight response, but this also causes your intestinal muscles to contract—that's what causes stomach butterflies or a sudden need to use the bathroom. In modern life, the threat is more likely to be a big sales presentation than a cave bear (though your boss may remind you of a cave bear, I get it), but stress is stress, and your digestion can suffer. For now, know that stress can affect the speed at which food moves through your digestive tract, make your gut more permeable, and affect your immune function. (Stress has such a powerful effect on your digestion that I have to mention it throughout this book. I'll discuss it in depth in chapter 7, "The Zen Corner.")

Another pathway that links the gut and the brain is a group of neurotransmitters, the most important of which is serotonin. Throughout your body, neurotransmitters are the natural chemicals that transmit impulses from one neuron (nerve cell) to another. That could be between neurons in your brain and central nervous system, or in your muscles, or in your gut. In the brain, serotonin plays a key role in mood regulation and memory. It's known as the happiness hormone, so we hear a lot about serotonin's role in depression. Selective serotonin reuptake inhibitors (SSRI) drugs such as escitalopram (Lexapro) and fluoxetine (Prozac) help depression symptoms by increasing levels of serotonin in the brain.

The reality is that most of the serotonin in your body—about 90 percent—is produced in your gut, not your brain. It's made by your handy general contractor: the microbiome. Here's another example of why it's so important to have a diversity of bacteria in your gut. Not enough of the kind that stimulate peripheral serotonin production can lead to a shortage that causes digestive problems—and may also cause depression.[1]

Most of the serotonin in your brain originates from your food. The process starts when you eat foods containing the essential amino acid tryptophan. These are pretty much any animal food, like eggs, milk, cheese, fish, meat, and poultry. Some plant foods, such as chocolate (yes!), bananas, pineapple, nuts, tofu, and spinach are also good sources of tryptophan. In the gut, some of the tryptophan gets converted to something called 5-hydroxytryptophan (5-HTP) and enters your circulation. When 5-HTP reaches your brain, it gets converted into serotonin. The process explains why turkey is notorious for putting you to sleep after a big holiday meal. Turkey is an excellent source of tryptophan, which means that about half an hour after you eat a plateful of it, your brain gets a big dose of serotonin, the feel-good-and-relax neurotransmitter. You're asleep on the couch before you know it.

Gut serotonin mostly stays in your gut, letting you know when you're full, regulating fluid levels, moving food through your digestive tract, sensing pain and nausea, and allowing for normal bowel function. Serotonin also helps protect you against dangerous toxins from your food. If you eat something toxic, serotonin makes your digestion speed up, so the food gets expelled quickly. Some of the serotonin gets carried into the circulation, where it plays a role in your bone health and wound healing. It's also theorized that some of that circulating serotonin reaches your brain. That suggests a way to treat depression without drugs by manipulating your gut microbiome to make more serotonin, which would then move from your gut to your brain and help make up for any deficiency. The idea that you could take a probiotic instead of an SSRI antidepressant is actively being studied now.

And while we're on the subject of antidepressants, sometimes they're used to treat the common gastrointestinal condition irritable bowel syndrome (IBS). Some of my patients with IBS find that their symptoms are calmed greatly by taking a low dose of an antidepressant. The doses I prescribe for these patients are lower than what would be used for depression, which suggests they have a more local neurotransmitter effect on the gut rather than on the brain—the drugs work by acting on the ENS in the same way they act in the brain. Some patients also benefit from psychological counseling. Again, the goal is to improve their gut symptoms by improving the gut–brain communication. By helping their big brain cope better with their symptoms, counseling often impacts their second brain, reducing symptoms and improving their depression. Talk about a virtuous cycle!

The next arm of the gut–brain axis is your immune system. When your immune system gets activated for whatever reason, cytokines—chemical messengers that coordinate the activity of your immune cells—kick off inflammation. When cytokines reach the brain, they can have a powerful impact on the neurotransmitters, affecting how they're made, released, and used. In other words, cytokines that originate in your gut can end up in your brain, where they can cause you to feel lousy. You get tired, withdrawn, lose your appetite, can't think straight. Your body is telling you to rest and direct your energy toward getting better, which of course makes sense if you're sick. But if you have chronic, low-level inflammation, those cytokines in your brain may consistently trigger your body to feel like this *all the time*, which is certainly no way to live.

The final pathway for gut–brain communication is bacterial metabolites, also produced by your gut bacteria. Some bacterial metabolites can end up in the circulation and get carried around to your entire body—including your brain. To keep unwanted molecules in the circulation from doing damage, you have a blood-brain barrier (BBB). The walls that form the blood vessels in the brain are lined with tightly spaced cells that block most large molecules from entering the brain. Does that sound sort of familiar? It's similar to the lining of your small intestine. And just as the tight junctions of your small intestine can open too widely and become permeable, so can the blood-brain barrier. That can let bad molecules enter your brain, which down the line causes the inflammation and other damage we associate with aging and disease.

The same poor diet that causes leaky gut can damage the BBB—and a healthy diet can be protective. To take one example, bacteria in a healthy gut produce the short-chain fatty acid butyrate, which keeps the lining of the colon strong. Butyrate is also important for the integrity of the blood vessel walls in the blood-brain barrier. If the blood vessels are sturdy and strong, so is the barrier.[2]

By now, I hope you're convinced that your general contractor and your architect need to work very closely together to keep your home running smoothly. But what about when that relationship goes sour? Let's find out.

Depression and Your Microbiome

A lot of research in the area of bacterial metabolites is focused on the causes and treatment of major depressive disorder (MDD). People with MDD have some real differences in the composition of their gut microbiome, compared to people without MDD. They tend to have more bacteria from the Bacteroides family, for example, and fewer bacteria from some other large groups. It's possible that this alteration in the gut bacteria diversity and numbers is related to why people with MDD also tend to have high levels of inflammatory cytokines and more overall systemic inflammation. These and other discoveries are very exciting, because they point toward ways to accurately diagnose severe depression and nondrug interventions, like dietary changes and probiotics, that might help.[3]

Migraines and the Microbiome

The intense, throbbing pain of a migraine headache is surprisingly common. About 18 percent of American women get these awful headaches (not to mention nausea and sensitivity to light!) about once or twice a month, and acute migraine attacks account for some 1.2 million visits to the emergency room every year.

A tendency to get migraines is probably genetic. For reasons that are still poorly understood, the headaches themselves have a wide range of triggers. Stress, anxiety, lack of sleep, hormonal changes in women, bright or flashing lights, and alcohol are just some. Some people get migraines from eating trigger foods. A common one is chocolate—so sad! Others can include aged cheeses, cured or processed meats, and foods with added MSG.

Many of the things that trigger a migraine can also cause or be caused by inflammation from dysbiosis and leaky gut syndrome. If that's the case, then probiotics to help resolve the gut issues should help, right? According to a recent study, they can. For the study, seventy-nine people who got migraines were randomly assigned to take a probiotic or a placebo daily for ten weeks, without knowing which they were getting. At the end of the ten weeks, the participants who took the probiotic supplement reported significantly fewer migraines than those who took the placebo. The probiotics group also needed fewer migraine drugs, had shorter attacks when they did have a headache, and lost fewer days to migraine symptoms than the placebo group.[4]

This study and others strongly suggest that the gut microbiome plays a role in triggering frequent migraines. Drugs to head off or treat a migraine have some unpleasant side effects, plus using some of them long-term can actually make the headaches worse. A better diet, some lifestyle changes, and ingesting probiotics daily (in food or supplements) can reduce the need for medications and improve the quality of life for some migraine sufferers.

Neurodegenerative Disease

When it comes to getting old, most people (including me) say their biggest fear is getting dementia or otherwise losing cognitive ability.

And what's especially scary is that the brain changes associated with dementia can predate symptoms by ten to twenty years. Research now shows that your microbiome seems to play an important role in causing degenerative neurologic diseases, such as Alzheimer's disease (AD), Parkinson's disease (PD), and possibly multiple sclerosis (MS).

Today some 5.8 million Americans are living with the memory-robbing effects of Alzheimer's disease. AD is mostly a disease of aging; of all the people with AD in the United States, about 80 percent are age seventy-five or older. The hallmark of AD is brain damage from a build-up of proteins called beta amyloid and tau. The proteins accumulate as plaques and tangles in the regions of the brain that control memory, causing irreparable damage.

AD has many contributing causes—some as yet unknown, and some known, including smoking, heavy alcohol consumption, and genetics. Your microbiome can also play a significant role. We know that the makeup of your gut bacteria changes as you age, with a tendency to become less diverse overall. The balance of species also shifts, with some becoming more numerous and others diminishing. It's possible that in people with AD, this shift creates dysbiosis and increased gut inflammation and permeability. Stool samples from people living with Alzheimer's disease have a less diverse community of gut microbes than samples from people the same age but without AD.[5] People with AD tend to have lower levels of *Bifidobacterium* and *Firmicutes*, both among the most common bacteria in a healthy gut. Researchers have also shown that the changes in the gut bacteria are linked to the amount of beta amyloid and tau proteins that appear in the spinal fluid.[6] People with AD also tend to have three times as much LPS (a toxic breakdown product from some gut bacteria) in their blood than people without AD—and we know that LPS can contribute to the damaging protein buildup in the brain.[7]

While the connection between the microbiome and major degenerative diseases is still being investigated, the good news is that it's never too late to start taking control of your health. By taking action to diversify your microbiome now, you could very well be taking care of your brain in the long term.

Use It or Lose It

As you age, your brain naturally loses some processing power—just like my so-called smartphone does as soon as the warranty runs out. You might find yourself forgetting the name of an acquaintance, or having trouble recalling a word now and then. You might not be able to concentrate on something as deeply or for as long as when you were twenty years younger. As a doctor hitting her fifties, I'm here to tell you that the minor memory and focus lapses that start to happen as you age are perfectly normal.

Here's a fun fact: your brain actually shrinks as you age—starting in your late thirties. Okay, fine, it's a *slightly* depressing fact. After that, with every subsequent decade your brain decreases in volume by approximately 5 percent. This shrinkage means you lose some nerve cells (neurons), which can affect memory, and you lose nerve receptors, which slow down your thinking a bit. But don't despair: the wisdom you've accrued over the years helps make up for these relatively minor losses. You can still do your job as well as ever, because your work experience easily compensates for sometimes forgetting the name of that new guy in sales. The changes aren't necessarily signs of impending dementia or Alzheimer's disease. But they are still annoying, and they are a good reminder that our inner architect needs a good diet and regular exercise to stay in top gear. We would certainly like to minimize the wear and tear on our brain if we can.

A healthy gut microbiome helps keep your brain fit and healthy as you age by keeping inflammation and its damaging consequences to a minimum. The best way to minimize inflammation is to prevent it—and that means eating a healthy diet. (I'll go into the details of that in the next chapter.) For now, I'll just say that what's good for the gut is good for the brain. Eat lots of fresh fruits and vegetables (especially greens), whole grains, and beans; get plenty of good fats (the omega-3 kind) from fish, nuts, and unprocessed vegetable oils; cut way back on red meat and processed meat like bacon; and go easy on alcohol. Most of all, avoid highly processed and manufactured foods. These products (I can't really call them foods) are full of sugar, salt, bad fats, and chemicals and are almost fiber-free—they're bad for you in every way.

Exercise also seems to counteract this phenomenon of brain-shrinking. A study of activity in elderly people (average age seventy-five) showed that the people who were more engaged in walking, gardening, swimming, or dancing had larger brains on MRIs than their sedentary counterparts. How exercise boosts brain health involves two main factors. It increases your cardiovascular strength, which then allows your heart to deliver more oxygen and nutrients to your brain, and it also positively affects your microbiome.[8] I'll go into this more in chapter 6, "The Home Gym."

There's another important dimension to brain health: using it. Being intellectually and socially active keeps you mentally alert and engaged. It supports a positive mood and helps protect the brain from cognitive decline. As part of your Gut Renovation, I strongly recommend reading more—it's a great way to exercise your brain and expand your horizons. In fact, you're already racking up brain health points just by reading this book. Go you! You could even switch careers or make bigger life changes (I became an entrepreneur at age forty-two!), but there are less dramatic ways to exercise your brain. This could be as simple as a thirty-second switch-up, like using your nondominant hand to brush your teeth (warning: this can get messy at first). Or you can do crossword puzzles, learn a new language, take up a craft, play a musical instrument—anything you enjoy and do consistently will help keep your inner architect sharp and creative.

Your inner architect also needs some time off now and then to recharge. Sometimes doing nothing in particular—what neuroscientists call nontime—lets your conscious brain relax and gives your subconscious brain a chance to work. When you're out for a stroll, weeding the garden, or relaxing with a cup of tea is often when good ideas and solutions to problems come to you. I know this seems hard to carve out given the crazy schedules many of us have these days, but you can find some pockets of time if you make it a priority. Having plenty of unstructured time worked well for Albert Einstein, Steve Jobs, and many others—give it a try.

While alone time is crucial, one thing I can't recommend highly enough is staying connected to family, friends, and community. Over a lifetime, people with strong social relationships are 50 percent more likely to reach a healthy old age than people who are isolated. In other words, too much alone time can harm your health—it's as bad as smoking fifteen cigarettes a day.[9] People who feel lonely and isolated show increased blood markers for inflammation, probably because they're so stressed. And as you know by now, anything that raises your stress hormones and causes systemic inflammation also causes changes in your gut microbiome. We also know that increased social isolation in older adults can increase the risk of dementia by up to 50 percent.[10]

On the flip side, being married reduces your risk of dementia.[11] Interestingly, long-term intimacy with a spouse or life partner can also improve the gut health of both people. Your microbiomes sync up and resemble each other, with greater similarity among the range of bacteria families. You know how they say married people start to look like each other? Well, apparently so do their guts! Compared to their siblings or to people who live alone, partnered people also have richer, more diverse gut microbiomes.[12]

Staying socially active became much harder to do during COVID, but we have to be careful that social distancing doesn't become social isolation. I know Zoom calls with family and friends have gotten old, but don't give up on them! If you're feeling lonesome, reach out to your community of family, friends, and neighbors. In fact, thanks to technology like Zoom and FaceTime, it's actually easier to socialize (albeit virtually) with people who live far away. Another great way to stay connected is to volunteer in your community. And that endorphin surge you get from helping people (what's known as a "helper's high") helps your brain too![13] Remember also to pay it forward by reaching out to someone you know is feeling lonely. Sometimes a friendly check-in phone call can really help.

And speaking of being friendly, now that your architect and general contractor are in good shape and are working well together like the best of friends, what's next? When you do a Gut Renovation, where do you start? The heart of the home—the kitchen. Yummy! Let's go!

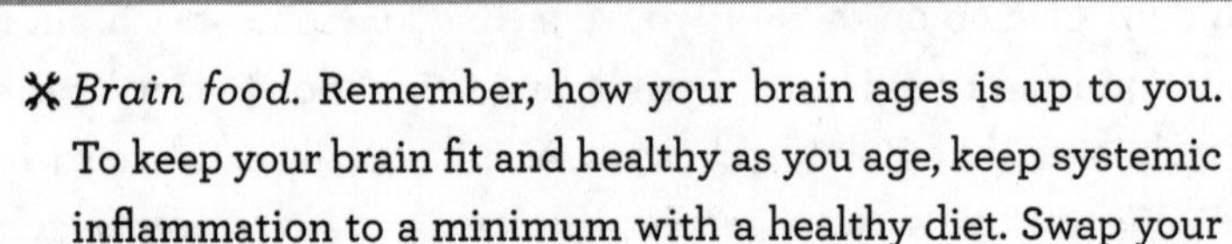

GUT TOOLBOX

- *Brain food.* Remember, how your brain ages is up to you. To keep your brain fit and healthy as you age, keep systemic inflammation to a minimum with a healthy diet. Swap your daily glass of vino for a bunch of red grapes and a juice shot made from leafy greens.

- *Mix it up.* Exercise helps counteract brain shrinkage as you age. Try an app like Class Pass where you pay monthly for unlimited classes of different styles. You may be salsa dancing one day and kickboxing the next—learning the new routines is the perfect way to work out your brain.

- *Think outside the box.* Challenge your brain actively every day. Join a book club—or start one. Volunteer for something you care about in your community. Use an app to learn a new language or sign up for a MOOC—a massive open online course. MOOCs are free, convenient, and open to everyone (more info at mooc.org).

- *Press pause.* Make time to relax and let your brain recharge. Download a meditation app and start with ten minutes a day. When your subconscious mind takes over, you'll find new ideas and solutions to problems.

CHAPTER 3

THE KITCHEN: EATING YOUR WAY TO A YOUNGER BODY AND MIND

The starting point for your Gut Renovation is the kitchen—you need to tear out your old eating habits and install new and updated ones. Think of it this way: When you update your kitchen appliances, they not only make your kitchen more functional, but they also make the space more pleasant visually. That's how it is with improving your diet. You'll look and feel better because you're making eating well a priority, something many Americans struggle to accomplish.

The typical American diet—high in fat, sugar, and processed foods—is wreaking havoc on your health. Your new approach to eating means better nutrition, a healthier gut, and maybe even an end to some digestive problems. And the best part is that the foods in this diet actually taste great. One more thing: although this isn't a weight-loss book, the diet you learn here (with tons of delicious recipes in the back of the book) will lead to a leaner you, along with better skin, hair, muscles—some of the awesome perks of a healthy gut!

What Should I Eat?

I spend a lot of time discussing diet with my patients, because what they eat can have a direct impact on their symptoms and how quickly they get better. My goal is always to help my patients make the dietary changes that will keep their digestion healthy. As much as I love seeing my patients, what I love even more is helping them get better so they can stay out of my office!

For almost all of us, the best place to start to optimize gut health is to take a virtual trip (or a real one if you can swing it!) to one of my favorite regions on the planet: the Mediterranean. Not only will you find gorgeous beaches and charming villages there, but you'll also find a superhealthy mode of living. The way of eating followed in traditional communities in the sixteen countries that border the Mediterranean Sea form the fundamental concepts of the Mediterranean diet. People who follow this dietary pattern, such as those living on the islands of Crete and Sardinia, tend to lead long, healthy lives that are largely free of the chronic diseases I see so often in my practice.

The Mediterranean diet is mostly plant-based; it's full of vegetables, whole grains, beans, nuts, and seeds, along with good fats and healthy proteins. Dairy products, fish, eggs, and poultry are eaten in low to moderate amounts; red meat is eaten very rarely. Dessert is usually fruit, not a high-calorie sweet dish. Added sugar, animal fats, tropical oils, and processed foods just aren't part of the diet.

One thing I love about this diet is the amount of scientific data behind it—which isn't the case with so many fad diets you may hear about. Numerous studies have shown that the basic Mediterranean diet helps prevent heart disease and stroke and reduces risk factors such as high cholesterol and high blood pressure. It's also very helpful for people with prediabetes and type 2 diabetes as a way to get their blood sugar down.[1] And for people at risk for Alzheimer's disease, a very recent study showed that the Mediterranean diet was protective against memory decline and brain shrinkage.[2]

The Mediterranean approach means you eat more servings of fruits and vegetables than the average American. Even though a standard serving is just half a cup, most Americans eat only one serving of fruit and one and a half servings of vegetables a day. That works out to some morning OJ and the lettuce, tomato, pickle, and little tub of coleslaw that come with a burger. (Sorry, but the ketchup doesn't count.) When you raise your intake to the recommended five servings a day, the difference is dramatic. Your overall risk of death compared to eating two servings a day drops by 13 percent. Specifically, you have a 10 percent lower risk of death from cancer, a 12 percent lower risk from cardiovascular disease, and a 35 percent lower risk from respiratory disease. Recent research suggests that eating more than five servings a day doesn't lower your risk of death any further, although it certainly has other health benefits.[3]

MIND Your Diet

The Mediterranean diet is an excellent starting point for healthier eating, but we can actually take the diet a step beyond and further reduce inflammaging with the MIND diet.

Imagine the Mediterranean diet with some tweaks that make it even better for you, including some that make it even healthier for your brain. That's the Mediterranean–DASH Intervention for Neurodegenerative Delay, called the MIND diet for short. The DASH part comes from Dietary Approaches to Stop Hypertension, an eating plan based on research sponsored by the National Heart, Lung, and Blood Institute (NHLBI). Because it emphasizes a plant-based diet and cuts your salt intake, DASH has been shown to lower high blood pressure and improve cholesterol, which of course reduces your risk of heart disease.[4]

The MIND diet takes DASH to the next level by focusing on brain health. It was developed by researchers sponsored by the National

Institute on Aging. The first major study published in 2015 showed amazing results. People who really stuck to the diet lowered their risk of Alzheimer's disease by as much as 53 percent; those who followed it only moderately well still lowered their risk by about 35 percent.[5]

Anything that lowers the risk of getting a serious condition like Alzheimer's definitely grabs my attention. So will any diet that's easy and tasty, and has a lot of variety. The MIND diet builds on the basic Mediterranean diet, but it also specifically eliminates foods that are bad for the brain and emphasizes certain brain-healthy foods like berries. What makes the MIND diet even more attractive to me is that many of the same foods that are good for your brain and circulation are also optimal for your gut.

While Mediterranean and MIND are both great diets, they weren't created specifically for gut health. In creating the Gut Reno diet, I used them as building blocks and added components that are uniquely tailored to boosting gut health. Think prebiotics, probiotics, antioxidants, and more, which all lead to better aging, more energy, and a lower risk of chronic disease. I lay out the Gut Reno diet in detail in chapter 11, but read on to learn about some of the main components.

Diet and Your Microbiome

The composition of your gut microbiome is shaped by what you eat. Sure, other factors like stress and genetics play a role, but study after study shows that diet is key. By eating a varied, nutrient-dense diet with plenty of whole foods, you develop a healthy microbiome full of the bacteria associated with good health in every dimension. Feed your microbiome junk—processed or fried foods, refined grains, sugar, salt, food additives—and your bacteria shift to the ones that are associated with poorer overall health, including heart disease and diabetes.[6]

That's what makes the basic Mediterranean eating pattern such a good dietary foundation. When you switch from the standard American diet, rich in meat, cheese, and dairy products, to a diet with much less meat and much more fiber, your gut bacteria change—for the better. The meat-heavy diet promotes the growth of unhealthy bacteria that are implicated in causing inflammation, including in the colon (colitis). When you move to a more plant-based diet with less red meat, your gut bacteria respond by increasing the number of bacteria that help break down fiber and reducing the number of potentially harmful bacteria. The change happens within a few days, which means quick results.[7] By reducing the amount of red meat in your diet, your bacteria will also be doing your heart a favor. When you digest red meat, one of the by-products formed by your gut bacteria is something called TMAO (trimethylamine N-oxide), which is known to increase your risk of heart disease. Researchers have shown that changing your diet to eliminate red meat can significantly reduce the level of TMAO in your blood—and reduce your risk of atherosclerosis, hypertension, and heart failure.[8]

The Mediterranean approach has also been shown to improve the overall health of older people. An important study in 2020 looked at how eating a Mediterranean-style diet for a year affected the health of 612 older adults who were at risk of frailty—which is age-related loss of strength and normal function, along with inflammation and cognitive decline. Frailty leads to a significant reduction in your quality of life and your life expectancy.

At the end of a year of eating the Mediterranean way, the lucky participants had larger numbers of bacteria families that are positively associated with lower frailty and less cognitive decline, and lower numbers of bacteria associated with inflammation. In other words, after a year of eating a delicious, varied diet, they were healthier overall and better able to avoid frailty in the future.[9]

When you move to a plant-forward, Mediterranean-style diet, you toss the sugar, the junk food, the processed foods, and most of the meat. That means tossing inflammaging and poor gut health. But what you eat instead matters even more. When you replace bad foods with more greens and other plant foods, you also increase your intake of three crucial dietary factors: fiber, phytonutrients, and fats (the good kind).

Fiber: Feed Your Flora

I've already mentioned the importance of fiber in your diet, and I'm going to mention it a lot more throughout this book. I love talking about fiber so much that by the end of this book you will have fiber coming out of your . . . well, you get the point. But what exactly is fiber? And why is it your BFF?

Dietary fiber is the parts of plant foods that can't be digested in the small intestine and get passed through to your colon. There are two varieties: soluble fiber and insoluble fiber, and both are essential for healthy digestion. Soluble fiber attracts water as it passes through the digestive tract and turns into a soft gel. Soluble fiber is found in beans, lentils, peas, barley, oatmeal, nuts, seeds, and some fruits (apples and peaches, for example). Insoluble fiber is made up mostly of cellulose: tough plant cell walls that don't absorb water. It's found in whole grains, nuts, and fruits and vegetables—think crunchy, stringy celery. It's the insoluble fiber in the plant material that helps keep food moving through the small intestine and adds bulk to your stool in the colon. That bulk gives your colon muscles a workout as they propel the stool forward. In this way, a high-fiber diet ensures that your digestion goes along smoothly and that you have regular, well-formed bowel movements.

Because you don't digest it in the small intestine, fiber doesn't add any calories to your intake. One more reason to love fiber! Once it reaches your colon, though, some of the fiber does get digested—by the trillions of bacteria in your colon. They use metabolic processes to ferment it, sort of like having your own internal sourdough starter. Fermentation converts the carbohydrates in the fiber into short-chain fatty acids (SCFAs), including one called butyrate. The cells lining your colon wall love butyrate. It's the fuel that keeps them healthy and active. SCFAs also inhibit the growth of some unfriendly bacteria, which gives the helpful ones more space in your crowded colon.[10] It's possible that SCFAs also protect you against intestinal inflammation and colorectal cancer through their effects on your immune system.[11] They may also play a role in appetite regulation and how your body produces energy. That's a really interesting area, because it could explain why some people struggle to keep their weight down even though they don't eat excess calories. Some people's gut bacteria may extract more calories from the exact same amount of food—so unfair! This is why it's so important to keep your gut well stocked with the right types of bacteria.

And it doesn't take long for fiber to effect positive changes in your microbiome. One study showed that in just two weeks on a high-fiber diet, the participants' gut microbiome composition significantly altered, with higher numbers of beneficial bacteria compared to before the study.[12]

If your diet is high in fiber, you're giving your good bacteria the environment they like best. At the same time, when you add more fiber to your diet, you're helping your health in many other ways:

- *Lowering cholesterol.* Fiber in your digestive tract helps reduce how much cholesterol you absorb from your food and lowers your body's natural cholesterol production.
- *Keeping your weight down.* When you substitute high-fiber plant foods for low-fiber processed foods, you take in fewer calories. High-fiber foods also slow your digestion, which can really help you feel full for longer.
- *Controlling your blood sugar.* Because high-fiber foods are digested more slowly, the carbohydrates in them are slower to enter your

bloodstream. That helps you keep your blood sugar on a more even keel, avoiding highs and lows (this is especially important for people with prediabetes or type 2 diabetes).

- *Reducing your risk of gastrointestinal cancer.* As a gastroenterologist who screens for cancer every day, I can't emphasize this enough. A diet high in fiber helps protect you against colon cancer and some other cancers, including breast cancer. It also helps protect against some other gastrointestinal diseases, like diverticulitis and irritable bowel syndrome.

How much fiber do you need to keep your gut bacteria happy? According to the National Academy of Sciences, a woman should get 25 grams of fiber a day; a man at least 30. Another way of looking at this is aiming for 14 grams of fiber for every thousand calories in your diet. The sad reality is that very few people actually get that much fiber. The average woman gets only 12 to 15 grams a day, while the average man gets only 16 to 18 grams. As I see every day in my office, not enough fiber is a contributing cause to a lot of digestive problems. I tell my patients that the guidelines are actually just the bare minimum for fiber. I recommend aiming higher, by at least 10 more grams a day, and make sure you get a mix of both soluble and insoluble fiber. The best way to do this is by eating a diet with minimal sugar, very few processed foods, and lots of plant-based foods. In other words, a plant-forward, Mediterranean-style diet. When you start your day with a bowl of oatmeal (14 grams of fiber per 1 cup serving), then for lunch have a bowl of vegetarian chili (on average, about 12 grams of fiber per 1 cup serving) and a small salad (on average, about 4 grams of fiber per 1 cup serving), you're already at nearly 30 grams—and you haven't even had dinner yet.

Want to add more dietary fiber in a way that's been proven to improve your microbiome—and also tastes great? Eat an avocado. One avocado has 6 grams of fiber, along with 7 grams of healthy monounsaturated fat. A recent study showed that eating an avocado every day increases the number of gut bacteria that are especially good at breaking down dietary fiber, while also increasing your overall gut microbiome diversity.[13] But feeding your microbiome is just one of the many benefits of eating avocados. They're also a great source of vitamins like folate, vitamin K, and vitamin E, as well as potassium, antioxidants, and healthy fats, which make them good for blood pressure, vision, arthritis, cholesterol, and more. Talk about getting a bang for your buck! Pass the guac, please!

A word of caution (so you don't send me angry emails): it's important to work up to more fiber in your diet *gradually*. Suddenly adding a lot of fiber can cause uncomfortable gas, bloating, and diarrhea. Slow and steady wins the race here. (I'll talk more about this in the next chapter, one of my favorites: "The Bathroom.")

Prebiotic and Probiotic Foods

When your diet is rich in natural probiotic and prebiotic foods, your microbiome feels the difference—and so do you. These foods are emerging as perhaps the most potent dietary weapons in the age-defying tool kit.

Prebiotic foods are high in the insoluble fiber that feeds your beneficial bacteria. The insoluble fiber goes by a number of hard-to-remember names: galactooligosaccharides, fructooligosaccharides, oligofructose, inulin, and chicory fiber. Since you're probably not equipped to do chemical analysis in your kitchen, look for these great prebiotic foods instead: almonds, apricots, asparagus, artichokes, beans, Jerusalem artichokes (sunchokes), garlic, mango, pears, pistachios, and watermelon. Foods that are high in antioxidant polyphenols are also good prebiotics—your bacteria love them. Good choices here are apples, berries, tea, dark chocolate, and flaxseed. In fact, between the insoluble fiber and the polyphenols, almost *any* fruit is a good prebiotic food.

Probiotic foods are foods that contain beneficial bacteria, usually in the *Lactobacillus* and *Bifidobacterium* families. Sauerkraut and kimchi, made from cabbage, are good sources; so are pickles. Miso and tempeh, both made from soybeans, are also rich in beneficial bacteria. Yogurt and kefir are made through lactic acid fermentation of milk, which makes them probiotic-rich; you can also get vegan and dairy-free versions. Look on the label for the words "live and active cultures" to be sure you're getting a product that contains plenty of live bacteria.

Although you'll get more than this in the Gut Reno diet, at a minimum you should aim to eat at least one portion daily of a high-prebiotic food and a daily serving of a probiotic-rich food. Supplements of both prebiotics and probiotics can also be added to ensure a steady supply.

Fight Damage with Phytonutrients

Phytonutrients is a sort of catch-all name for the thousands of natural chemicals found in plant foods. They're the substances that give these foods their characteristic taste, smell, and color. Phytonutrients make oranges orange, red peppers red, and blueberries blue. Plants make these chemicals to protect themselves from sun damage and to fend off things that want to eat them, like fungi, insects, and us. Fortunately for humans, some of the phytonutrients that protect plants also taste good—and pass their protection on to us.

Specifically, many phytonutrients are powerful antioxidants that protect your cells from free radical damage. The term *free radical* sounds like an anarchist group that wants to cause instability and chaos—and that's essentially what they do in your body. The thousands of chemical processes that go on in your body every second of every day create free radicals, the body's equivalent of exhaust fumes. Free radicals are highly reactive, short-lived molecules that are missing an electron.

That molecule is unstable and desperately wants to fill in the gap, so it will grab an electron from another nearby molecule. But that turns the next molecule into a free radical—and so on down the line. All those free radicals blundering around in your cells and grabbing electrons can do a lot of damage to the cell membrane, the internal cell structures, and even the DNA in the nucleus. What are the potential consequences of that damage? Inflammation, organ damage, and even cancer.

Fortunately, your body has powerful natural defenses against free radicals. Antioxidants, substances found inside and outside your cells, provide the missing electrons the free radicals crave. They quickly neutralize them and stop the cascade, without creating new free radicals in the process.

Your body manufactures antioxidants, but the antioxidants you get from your food are a bonus that gives you extra free radical protection. Vitamin C, which you can get only by eating it, is a superstar antioxidant. Your body can use the help, because your metabolism isn't the only way free radicals are created. The ultraviolet light in sunshine can do it, for example, which is why people who spend too much time in the sun are more likely to get skin cancer and cataracts. When you're creating free radicals faster than your own antioxidants can quench them, the additional antioxidants from your food can make a difference.

New Tools for Your Kitchen

As part of your kitchen Gut Renovation, I recommend a new coffeemaker, a teapot, an extra-big spice rack, a wine rack, and a freezer. Here's why:

Coffee. Don't you love the smell of fresh coffee in the morning? The delicious aroma comes from the many, many phytochemicals in coffee—not from the caffeine, which is odorless. The phytochemicals mostly function as antioxidants in your body. They're so effective that drinking just one cup of coffee a day is linked to a 3 to 4 percent overall lower risk of death. Moderate coffee drinkers (three to five 8-ounce cups a day) have less type 2 diabetes, heart disease, colon cancer, Parkinson's disease, and cognitive impairment. Coffee drinkers also have more diversity in their gut bacteria compared to non–coffee drinkers, and their microbiome tends to be higher in anti-inflammatory bacteria. The more coffee you drink (decaf included), the healthier your gut microbiome—even if your diet isn't that great otherwise.[14]

Green tea. Don't let the delicate flavor of green tea fool you—this stuff is full of phytonutrients. It's rich in antioxidant compounds, such as epigallocatechin-3-gallate (EGCG), that may help prevent colon cancer and other cancers.[15] Green tea is also high in L-theanine, an amino acid that can help curb anxiety and help improve cognitive function.

Spices. The spices you use in cooking get their flavorful qualities from a very complex mix of phytonutrients. Take cinnamon, for instance. That simple little spice we use for our apple pies actually contains eleven different phytonutrients that have antioxidant effects. Turmeric is another powerhouse antioxidant spice that has been shown to cause positive changes in the microbiome, providing greater bacterial diversity and decreasing dysbiosis.[16] The more varied your spice rack, the more antioxidants you'll get. All the more reason to use spices liberally in your cooking and experiment with new ones from around the world.[17]

Red wine. Fermented foods are good for you, right? Well, red wine is made from fermented grapes—and *in moderation* (no more than one glass a day), it's good for your gut. The alcohol from the fermentation process isn't what helps (so sadly, other types of alcohol don't offer this benefit). It's the specific polyphenols in red wine that can improve the bacterial diversity in your microbiome. Red wine in moderation may also help your heart, lower your cholesterol, and help keep you at a healthy weight. As little as one glass every two weeks is enough to improve bacterial diversity, but more than one glass a day doesn't make diversity even better.[18]

Freezer. A key goal for gut health is increasing your intake of fruits and vegetables to at least five (preferably more) half-cup servings a day. The cost of that can add up. It's a sad reality that potato chips are cheaper than regular potatoes. To keep your costs down and always have a good supply of fruits and veggies on hand, stock up on frozen foods. From a nutrition standpoint, they're actually just as valuable as fresh foods, because they're picked and frozen at the peak of ripeness. In fact, frozen kale has even more antioxidants than fresh kale, and frozen peaches have more vitamin C than fresh peaches. Freezing also lets you take advantage of seasonal abundance from your own garden or the local farmers market and gives you a variety of choices year-round. The drawback is that the texture of the fresh version is changed by freezing. You can easily work around this by using the frozen stuff in soups, stews, casseroles, smoothies, and other dishes where the texture doesn't matter much.

Pass on Processed Meats

There's only one food I ask all my patients to basically stop eating: processed meats. I know it's difficult to give up bacon, ham, deli meats, hot dogs, jerky, and sausage (no more pepperoni pizza!), but these meats are just really unhealthy. They're full of salt, bad fats, and chemical additives. The nitrates and nitrites that are added to processed meat to preserve them are known carcinogens that raise your risk of getting cancer, especially colon cancer. The World Health Organization classifies processed meats in general, not just their individual ingredients, as a known carcinogen. A serving of processed meats once in a blue moon won't kill you—I eat my annual hot dog on July 4—but trust me, it's best to keep the processed meats to the bare minimum.

Friendly Fats for Your Friendly Bacteria

Fat is bad for you, right? Not so! Bad fats are bad for you, but good fats are essential for your health. The difference lies in the structure of the fat molecules. Here's how it breaks down:

- *Saturated fats are fats that are solid at room temperature, like butter.* Most saturated fats come from animals (coconut oil is a rare exception). In general, saturated fats are bad for you because they raise your cholesterol and your risk of heart disease and stroke. Foods high in saturated fat can trigger colon contractions in people with bowel problems. They also affect your gut microbiome by increasing the number and type of inflammation-causing bacteria.
- *Monounsaturated fats are mostly plant oils, including olive oil and peanut oil.* When they're cold-pressed and minimally processed, they're good for you. Avocados are a fabulous source of mono fat—half an avocado has a whopping 12 grams. Monounsaturated fats help lower cholesterol and are good for your gut bacteria.
- *Polyunsaturated fats include some plant oils, like sunflower seed oil, and fish oil, also known as omega-3 fatty acid.* Fish oil is a great way to give your body the essential fat it needs to build cell membranes and make the neurotransmitters, hormones, enzymes, and other chemicals it needs to keep things humming along.
- *Trans fats are unsaturated vegetable oils that have been heavily processed (hydrogenated) to make them soft at room*

temperature—like margarine. Trans fats are also widely used in processed and manufactured foods. They're so bad for your arteries that the FDA makes food manufacturers list them on the label if the amount per serving is more than half a gram.

So, in summary: Mono and poly fats, good; your good gut bacteria like them. Saturated fats, bad; they promote the gut bacteria that drive inflammation. Trans fats, *really* bad; they shift gut bacteria to those linked to obesity.

Protein Power

When I recommend a plant-based diet to my patients, they often worry about getting enough protein. I understand that concern. After all, your body needs amino acids, the building blocks of protein, for normal growth and repair and to make all your enzymes, hormones, chemical messengers, antioxidants, and other protein-based chemicals, so you want to be sure you're getting enough. Trust me, on the Gut Reno diet you'll get plenty of protein. Many people think of protein as coming only from animal foods like meat, fish, milk, and eggs, but in fact, plant foods are great sources of protein. You can easily meet your daily protein needs without animal foods because your need actually isn't that high—it's far less than most of us get each day.

The current recommended daily value for protein is 50 grams, based on a 2,000-calorie diet. Let's put that in perspective: An egg gives you about 6 grams of protein, a slice of cheese gives you about 7 grams, a quarter-pound burger gives you about 14 grams, and a baked chicken leg gives you about 30 grams. In plant food terms: a cup of cooked beans has on average about 15 grams of protein; a cup of cooked quinoa has about 8 grams; an ounce of almonds has about 6 grams. Even greens have protein: a cup of cooked kale has about 3 grams. So, if you follow my eating plan by eliminating red meat and cutting way back on other animal-based foods, you'll easily meet your daily protein needs while adding more variety to your diet. At the same time, you'll be improving your digestive health, because your gut really prefers plant protein and lean animal protein like fish and poultry over the saturated fats found in beef. Low-fat proteins are easier to digest and better for your gut bacteria.

MAGIC MUSHROOMS

I am a huge fan of mushrooms. Before you get too excited, I'm talking about the regular kind, not shrooms. Mushrooms are a hearty and delicious way to satiate your hunger—and they also happen to be incredibly good for you. They have a rich mix of vitamins and minerals, as well as a plenty of polyphenols, which give them their potent antioxidant activity. It's no wonder that several studies have demonstrated mushrooms' anticancer, antihypertensive, antidiabetic, anti-allergic, and anti-cholesterol activity. Recently it has become evident that many of these effects may be modulated by mushrooms' prebiotic influence on the gut microbiome.[19] So make sure you incorporate these gut-friendly fungi into your diet as often as possible. They may not give you wild visions like their psychedelic sisters, but they are still pretty magical!

Sugar: The Sweet Temptation

As part of your Gut Renovation, you should throw out the sugar bowl. A diet high in sugar means an increased risk of weight gain, heart disease, type 2 diabetes, cancer, fatty liver disease, cognitive decline, and even depression. All these bad outcomes occur because a high-sugar, low-fiber diet upsets your gut bacteria, which causes chronic inflammation in your gut.

Does that mean you can't ever indulge? No. Sweet treats like cakes, cookies, brownies, danishes, and donuts are fine for a special occasion (that's why they're a "treat") but they shouldn't be part of your daily diet. Another danger is processed foods. These are often loaded with added sugar, often in the guise of high-fructose corn syrup. While most fruit juices, soft drinks, and soda are little more than high-fructose corn syrup, water, and chemical flavorings, even packaged soups and salad dressings can have a lot of added sugar. By moving away from these foods and beverages, you're automatically cutting your sugar consumption.

Okay, so you decide to cut back on sugar by switching to artificial sweeteners instead—you swap your sugary soda for a zero-calorie version. Are you good then? Not exactly. You may save some calories, but at a possible cost to your gut bacteria. Some studies show that sweeteners, including aspartame (Splenda) and saccharin (Sweet'N Low), can alter the balance in your microbiome, reducing the number of beneficial bacteria.[20]

I confess I have a sweet tooth—many sweet teeth, if I'm being honest. Nowadays I satisfy my cravings for sweets with fruit, fresh or dried (without added sugar). The natural sugar in fruit—also known as fructose—is much healthier. You can overdo fruit sugar also, but it's a lot harder. Think of eating six chocolate chip cookies in a row—all too easy to eat them all and a couple more, right? Now think about eating six oranges in a row. Even if you managed to get them all down, the total calories would be much lower, plus you'd be getting good stuff like fiber, vitamin C, and potassium, instead of white flour, high-fructose corn syrup, and bad fats. I'll give you some more Gut Reno sweet treat options in chapter 11 and the recipes at the back of this book—something to look forward to!

New Kitchen, New Eating

As you rip out those old kitchen cabinets, rip out some of the foods they contain as well. Toss the sugary breakfast cereals, the cookies, the chips, and all the other junky snacks and low-nutrient processed foods. Stock your sleek new cabinets with whole foods the Gut Reno diet way. Replace the cereals with oatmeal; the cookies with dried fruit and nuts; the potato chips with pita chips. I know the changes will take some getting used to, and you're going to hear complaints from family members who want their favorite junk foods back, but hey, it's your house, and your renovation, right?

Weight Loss

To me, good health and optimal aging is about a lot more than the number on the scale. Still, we can't ignore the fact that being overweight or obese has a negative impact on your gut. In fact, having excess belly fat—a different kind of "gut"—is particularly bad for your gut health. The opposite is also true: dysbiosis can lead to weight gain, poor eating habits, and obesity. And during the COVID pandemic in particular, people were snacking more mindlessly, *and* moving less than usual. So it makes total sense that significant weight gain was one of the results and that losing weight is top of mind for many right now. The good news is that if you follow my Gut Reno eating plan in chapter 11, there's a good chance you'll lose weight if you need to, because you'll be eating a much healthier diet overall.

What I do think is important to understand is the way your weight affects your microbiome. Many overweight people have an imbalance between the two main families of bacteria in the gut, *Bacteroidetes* and *Firmicutes*, with the balance tipping toward *Bacteroidetes*. That can set off a cascade of metabolic changes leading to obesity—or do other factors that lead to obesity make the gut bacteria get out of balance? Would changing your gut bacteria profile change your weight?[21] That's a good question, because animal studies suggest it might.[22] Interestingly, eating a better diet helps your health regardless of weight loss. When overweight people switch over to a Mediterranean diet, they show improvements in their cholesterol levels and gut microbiome even if they don't lose weight.[23]

When I think about the way the gut microbiome affects our body, I sometimes wonder who's in charge: us or them? When it comes to food cravings and even addiction, it could be your microbiome is secretly running the show. Some of your gut bacteria have very distinct preferences for what they like to eat. The beneficial *Bacteroidetes* bacteria prefer to dine on fat, for example. If they get enough of it, they thrive and become more numerous. What if they're not getting enough fat from your diet to thrive? Or if they became so numerous they needed more fat than ever? They would want you to eat more fat. These thugs might make that happen by producing metabolites that would affect your brain and make you crave fatty foods. (Check back to chapter 2 for more on how your brain and gut interact.) A fair amount of animal evidence shows that gut bacteria can influence behavior, so it's very possible that food cravings and addictive eating originate with your bacteria.[24]

Special Diets for Gut Conditions

So far we've talked about general eating principles of healthy eating, but these may not work well for everyone. My patients with digestive issues often need to modify the basic approach to help them manage specific gut conditions. People with celiac disease, for example, can't eat foods that contain gluten, a protein found in wheat, barley, and rye. For celiac patients, gluten triggers an immune reaction that inflames and damages the small intestine.

Currently, the only treatment is a diet that strictly excludes gluten. Fortunately, with some modification, you can still easily eat the Gut Reno diet when you're gluten-free.

Lactose intolerance is another gut problem that's pretty easy to help through dietary changes. People with lactose intolerance no longer produce the enzyme lactase, which you need to digest the sugar lactose, found in milk and milk products like ice cream. Babies make lactase so they can digest milk, but most people around the world stop making the enzyme in large amounts after childhood. People of western or northern European descent may continue to make the enzyme as adults, but about thirty million Americans have lactose intolerance to some degree by age twenty. When they consume milk or milk products, they get bloating, gas, and sometimes diarrhea. If you have lactose intolerance, you can simply avoid dairy products, but who wants to go through life without the occasional ice cream? You can temporarily provide the missing enzyme by taking over-the-counter lactase supplements when you eat dairy. Some people with mild lactose intolerance can eat small amounts of milk-based foods as part of a larger meal. But if you have severe lactose intolerance, avoiding it is the only solution. Lactose is often used as a filler in processed foods, so read the ingredients label carefully—or better yet, don't eat processed foods.

Because the Gut Reno eating approach doesn't include many dairy foods, it's fairly easy to stick to it if you have lactose intolerance. One food we use a lot in the Gut Reno plan is considered a dairy product: yogurt. But guess what? Many people with lactose intolerance can tolerate yogurt. That's thanks to the bacteria in the yogurt. They actually digest the lactose, so by the time you eat it, not enough lactose to cause problems is still present. Bacteria to the rescue once again!

Bacteria may even get rid of your lactose intolerance problem for you. Some people who don't produce lactase can still tolerate milk because they have a lot of lactose-loving bacteria in their colon. The bacteria do what the missing enzyme can't—they digest the lactose and give off lactic acid, not hydrogen gas, as a by-product, so you won't feel gassy. It's possible that taking a prebiotic containing a non-digestible, complex sugar similar to lactose could feed the lactic acid bacteria in the colon and shift the gut microbiome enough to improve lactose tolerance. Experiments show this works at least some of the time and is helpful for reducing abdominal pain from eating lactose. Clinical trials are ongoing and will tell us more in the next few years.[25] Genetically engineered probiotics designed to produce the lactase enzyme in your gut is another promising approach that may be ready for prime time soon![26]

Another condition I see all the time in my office is heartburn, or the more severe form, gastroesophageal reflux disease (GERD). It's one of the most common reasons people visit a gastroenterologist; some estimate that 20 percent of the U.S. adult population suffers from it regularly. So what exactly is going on here? Well, we normally have hydrochloric acid in our stomach to kick off the digestion process. When there's too much acid, it can rise up into the esophagus, the tube that connects your throat to your stomach, and cause painful heartburn and inflammation. The lower esophageal sphincter (LES), a circular, muscular valve between the esophagus and the stomach, opens to let food into the stomach and then closes to keep stomach acid out of the esophagus. Acid reflux happens if the LES loosens or relaxes and lets stomach acid travel upward into the esophagus. Just about everyone gets heartburn occasionally, especially if you lie down after a big meal. If it happens to you a lot, you might have GERD, which is not only painful but can irritate the esophagus so much that it gets inflamed, swollen, or even ulcerated. Untreated GERD can lead to stricture, where the esophagus gets so narrow you have trouble swallowing and can cause changes that can lead to cancer.

Some small modifications can do a lot to help with heartburn and GERD. Wait at least two hours after you eat dinner before going to bed and stay vertical (sitting up or standing). When you're upright, gravity helps keep the acid in your stomach. Lie down on a full stomach and it's much easier for the LES to open and let the acid escape. By waiting, you give your stomach time to empty, so there's less pressure on your LES when you finally do lie down. Eating smaller, more frequent meals instead of three big meals a day can also help.

One of the most important factors in controlling acid reflux is a low-acid diet. The idea is that the foods on this diet keep your stomach's production of acid at a minimum, which in turn keeps it from causing heartburn or reflux. You also need to avoid foods that lower the pressure on the LES. Trigger foods for reflux are usually high-fat foods like french fries and bacon, high-acid foods like tomatoes and citrus fruits, caffeine, chocolate, peppermint, and spicy foods. Low-acid or alkaline foods are vegetables, noncitrus fruits, lean meats and seafood, grains like oatmeal and rice, and healthy fats. The Gut Reno plan has plenty of these to choose from.

When diet and lifestyle changes aren't enough to control GERD symptoms, over-the-counter and prescription antacid medications are usually the next step. Some early research suggests that probiotic supplements may also be helpful and may actually be a safer next step for treatment instead of antacids.[27]

Another diet I often recommend is the low FODMAPs diet for my patients with irritable bowel syndrome (IBS), because it can dramatically improve their symptoms. FODMAPs stands for "fermentable oligosaccharides, disaccharides, monosaccharides, and polyols." These are complex sugars found in milk, yogurt, and ice cream; wheat-based foods such as bread and breakfast cereal; beans; some vegetables, such as onions, garlic, and asparagus; and some fruits, especially apples, peaches, cherries, and pears. High-fructose corn syrup and artificial sweeteners are also on the avoid list. Do these foods sound familiar? I've already mentioned them in this chapter when talking about gluten, lactose, and fiber. People with IBS have more difficulty than usual digesting these foods because they can't break them down well in the small intestine. When the food gets to the colon, it tends to absorb a lot of water; it also provides a real feast of partially digested food to your bacteria. The result is bloating, cramping, gas, diarrhea, constipation, and possibly dysbiosis. To cut back on the symptoms long enough to help the bowel irritation get better, my patients follow a diet that's low in FODMAPs but still high in good nutrition.

The Dining Room

The final vital part of your kitchen renovation is learning *how* to eat. In the rushed modern world, taking the time to appreciate our food is harder than ever—and also more important than ever. If you live in a city like me (shout out to NYC!), you may not have a formal dining room in your home, but you can still make your eating area a space where you take the time to enjoy your meals and practice mindful eating.

Growing up in a Buddhist home, I was exposed to the concept of mindfulness at a young age, but I had no idea how much of an impact it could have on my physical and mental well-being until I was an adult. I truly believe mindfulness is one of the keys to healthy living. (You'll hear a lot more on this in chapter 7, "The Zen Corner.")

Mindful eating means making conscious decisions about how, when, and what you eat, ultimately leading to healthier habits. This means eating slowly, sitting down, in a relaxed setting where you can focus on the food—so no desk lunches and no multitasking (and I am the queen of multitasking, so I know the sacrifice I am asking)! Throw away the habits of eating standing up, or while watching TV, or gobbling up your food as quickly as possible so you can get back to other activities. Have I done all those things? Of course. Especially as a swamped medical resident, I ate as quickly and as mindlessly as possible. But now that I'm older and hopefully wiser, I realize that my body needs a different kind of eating pattern to thrive.

Another important component of mindful eating is listening to your body's cues of hunger and fullness. Before you start eating, ask yourself: Am I actually hungry? Or am I really just bored/anxious/sad? Sometimes we even confuse thirst for hunger. You should also consider how the food you are about to eat is going to serve you. Is this meal going to provide valuable nutrients that will give you strength and energy? Or are you going to feel uncomfortable and overfull afterward? Taking a moment to reflect and ask these questions before you shovel in the food can lead to better choices.

When you do start to eat, slowly chewing and savoring the flavors and textures of your food all help to control portions and appetite. Check in with yourself at intervals during the meal. Am I full? Why am I still eating? "Because it's there" might have been a good reason for the guy who first climbed Mount Everest, but it's not a good reason to clean your plate if you're satiated halfway through the meal.

Mindful eating also applies to food shopping. Before you buy food, consider your mood and how it may be influencing you. Ask yourself, am I buying these snacks because they're healthy and satiating, or am I buying Cheetos because I'm bummed out? You should also focus when you shop on the foods you specifically went to buy. This means no aimless browsing down the aisles to see what grabs your attention. And of course, never go to the market hungry!

Finally, there's a Japanese custom I love, one that's very similar to saying grace: Expressing respect and gratitude for your meal before you start. Take this moment to reflect on where your food is coming from, the people and the environment that produced it, and the people you're sharing it with—it will help you see more clearly how the food is impacting *your* environment—your body!

Now that you've renovated your kitchen, it's time to turn your attention to a much smaller, but equally important space—your bathroom.

GUT TOOLBOX

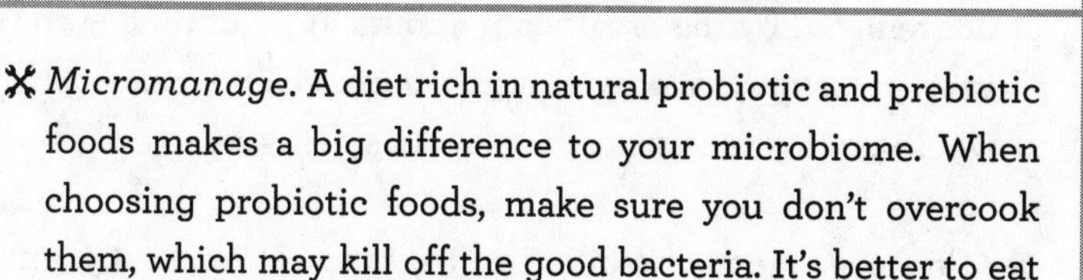

- *Micromanage.* A diet rich in natural probiotic and prebiotic foods makes a big difference to your microbiome. When choosing probiotic foods, make sure you don't overcook them, which may kill off the good bacteria. It's better to eat them uncooked or added only toward the end of the cooking process.

- *Focus on fiber.* Dietary fiber—at least 25 grams a day—is the key to improving gut health. Stir 2 tablespoons of chia seeds into your morning yogurt or smoothie for an easy 11-gram fiber boost. Tuck a piece of fruit, some carrot sticks, or a high-fiber snack bar into your bag for a quick bite, order a side salad when eating out, munch on popcorn or high-fiber veggie chips while you watch TV.

- *Follow the rainbow.* Those fiber-filled fruits and veggies are also great sources of phytonutrients that provide important health benefits. Aim for at least one colorful fruit (apple, plum, orange, kiwi, berries) and two colorful veggies (tomato, carrot, kale) a day as part of your five-a-day goal.

- *Brush it off.* When I'm stressed, I find myself wanting to endlessly graze on snacks, foraging through cupboards for something crunchy or sweet. One trick I use to stop myself is to brush my teeth immediately after I finish a meal. Once I do this, I don't feel like eating again (my teeth are so clean!) and I also know nothing tastes great when it is mixed with that lingering bit of toothpaste.

✖ *Work it out.* Work stress or boredom can lead to desk-side munching. Keep healthy snacks on hand at work so you're not tempted by the breakroom donuts. A jar of nuts is convenient; so are dried fruit, yogurt cups, healthy snack bars, whole-grain pretzels, and low-fat string cheese.

✖ *Mind over matter.* While the goal is to eat mindfully, this isn't always possible. But if you have a clean kitchen (no junk food!), you'll be less likely to mindlessly grab an unhealthy snack. So, do a biweekly pantry purge to make sure you (in a weak moment) or your kids haven't sneaked in any highly processed foods.

down and digested in your small intestine or further broken down by bacteria in your colon. Food waste makes up about 15 percent of your stool. Stool also contains dead cells from the gut lining, dead bacteria, some live bacteria (they're why stool smells), and some other metabolic wastes. The biggest component of your stool, however, is water. Even after fluid from your food and drink has been absorbed by your digestive tract, water still makes up about 75 percent of every bowel movement. In total, your bowel movements add up to at least four hundred pounds a year—talk about a load of crap!

Until the stool reaches your rectum, the process is automatic—meaning you're not aware of it. But when enough stool reaches your rectum, stretch receptors send you the message that it's time to go. To move smoothly, the defecation process depends on getting enough fiber in your diet to give the stool bulk and form and on having plenty of fluid in your diet to keep the stool soft and easy to pass.

Why am I regaling you with all of this information about stool? To impress upon you that you must look at your stool, people! There's a wealth of information in there, but you'll miss it if you just flush it down the toilet without taking a glance. Now, for some of you, this isn't news; you're already diligently checking out your stool. When I ask my patients to describe the appearance of their bowel movements, some proudly produce photos. Thanks to the smartphone, even amateurs have become professional-level stool photographers.

But other people shy away from this topic altogether and need a little encouragement, so here are some colors and textures to look out for. Your stool is usually a shade of brown because of bilirubin, a breakdown product from old red blood cells. But stool color can vary, mostly depending on what you've been eating. Beets and cranberries, for instance, can turn your stool temporarily red; spinach can make it seem almost black. So can iron supplements and pink bismuth (Pepto-Bismol). Green stool can come from bile pigment when you have diarrhea and food moves through your digestive tract too quickly for your gut bacteria to break it down to its normal brown color. It could also be from celebrating St. Patrick's Day too enthusiastically and consuming a lot of green food dye.

Many of these color changes are normal; the ones that concern me, though, are those that could indicate internal bleeding. Maroon-colored stool could mean intestinal bleeding from the lower part of the small intestine or the colon. Black stool that's also very stinky and tarry could mean internal bleeding from the stomach or the upper part of the small intestine. Bright red streaks on your stool usually mean bleeding from lower down in your colon. It could be hemorrhoids, an annoying problem but not one that's usually very worrisome. But any form of blood in the stool should always be evaluated by your doctor, as that's one symptom that could indicate cancer.

You'll also want to call your doctor if your stool is yellow, very smelly, bubbly, frothy, or hard to flush, because it could mean celiac disease or serious liver or pancreas disease. Another color to look out for? Pale, clay-colored stools could indicate gallbladder disease.

Constipation

One of the most common plumbing problems that can occur in your bathroom is the dreaded clogged pipe, aka constipation. This is one of the most common gastrointestinal complaints, sending about 2.5 million Americans to the doctor's office every year—and that's not counting the people who deal with it on their own. Roughly sixteen out of one hundred adults have symptoms of frequent constipation. It gets more common as you age, doubling in incidence after age sixty.[1]

Constipation is defined as:

- *Fewer than three bowel movements a week*
- *Stools that are hard, dry, or lumpy*
- *Stools that are difficult or painful to pass*
- *Feeling that not all stool has passed*

But as I tell my patients, you're only constipated if you feel constipated. Everyone is different and if your regular pattern of bowel movements isn't causing any pain or discomfort, then you're most likely fine, even if it technically means you're constipated. What is more concerning is any change in your pattern or if your pattern is causing you distress—that's when we need to investigate.

The most common cause of constipation is a diet that's too low in fiber. You need the fiber to hold water and give your stool bulk so it's easy to pass. If you start following my Gut Reno eating plan (see chapter 11), you'll be getting a lot more fiber in your diet. You'll notice the difference as your stool consistency, shape, and frequency improve. Being dehydrated can also cause constipation, which is one reason why water is so key to gut health.

Constipation can have a wide range of other causes, including many medical conditions. Pregnant women often get constipation; some drugs, such as opioid painkillers and antihistamines, or supplements like iron or calcium, can cause constipation. So can some bowel problems (I'll get into those later in this chapter). Stress can do it, as can being too sedentary, traveling into another time zone, and not getting enough exercise. Ignoring the urge to defecate can also cause constipation. A cause that's sometimes overlooked, especially in women, is an underactive thyroid gland. And sometimes you just have frequent constipation for unknown reasons—what we call functional or chronic idiopathic constipation (CIC).

TREATING CONSTIPATION

Many times you can fix constipation with some simple dietary and lifestyle changes. Stay away from junk food, which can slow down your digestion. Increase your fiber intake by eating more fruits and vegetables. Remember to increase fiber gradually, because suddenly adding a lot of fiber to your diet can cause cramps, bloating, and gassiness.

Also be sure to drink plenty of plain water—at least a liter a day—to avoid dehydration and to keep your colon from absorbing too much water, which leads to hard stools. You should also make an effort to get more exercise. This keeps the colon muscles healthy and improves the tone of the pelvic muscles you need to have a bowel movement. Because dysbiosis has been linked to constipation, taking prebiotic and probiotic supplements can also be helpful.[2]

When increasing dietary fiber intake isn't enough, it may be time to try fiber supplements, such as psyllium, methylcellulose, and/or calcium polycarbophil. The powder gets stirred into a glass of water and drunk. When it reaches the colon, the powder forms a soft gel that helps the stool hold more water and become bulkier and easier to pass. Since we know that fiber is generally great for your gut health, these supplements are usually safe to use over long periods of time if needed.

These steps are usually very helpful, but for chronic constipation we sometimes have to take stronger measures: laxatives. We usually start by trying nonprescription osmotic laxatives, which work by drawing water into the colon and stimulating a bowel movement. Nonprescription stimulant laxatives are a very distant second choice for constipation. To help patients with hard, difficult-to-pass stools, I sometimes recommend a stool softener such as docusate. These products can be useful for people who are recovering from surgery and for people who need to take opioid painkillers, which cause constipation. Finally, there are prescription medications for constipation that are generally reserved for severe cases.

COLON CLEANSES AND COLONICS

Colon cleansing and colonics may be trendy and have some celebrity cachet, but they don't offer any true benefit. Your colon doesn't need cleansing—it cleans itself every time you have a bowel movement. When you do a colon cleanse by eating whatever restrictive diet with magic ingredients the cleanse includes, you're basically just making yourself have a lot of extra bowel movements. When you have a colonic by having a tube inserted into your rectum and washing gallons of water through your colon, you're emptying your colon very completely in one big bowel movement. Next time you eat you're right back to forming stool, so you really haven't changed anything. Colonics are uncomfortable, expensive, and have the added thrill of possible infection from improperly sterilized equipment or potential damage to your colon lining. What's the worst thing about these procedures? They remove a lot of the good bacteria from your colon all at once.

Diarrhea

On the opposite end of the spectrum, we have the leaky pipe problem: diarrhea. Fortunately, most cases of acute diarrhea cause just a day or two of discomfort and you return to normal bowel movements quickly.

Acute diarrhea has many causes, but the most common ones are viral infection and food poisoning. Viral infections like flu, norovirus, and rotavirus (notorious in kids) are the usual cause of what we vaguely call stomach bugs or stomach flus. They're very contagious and almost unavoidable—you're likely to catch one every few years. Food poisoning is slightly different, because it's usually caused by eating or drinking something contaminated with one of the usual bacterial suspects, like salmonella or *E. coli*, or their toxins. For these types of episodes, waiting it out is usually all that's necessary, along with the DIY measures I'll describe later. Don't rush to antibiotics in this situation. They probably won't shorten the duration of your symptoms and they could disturb the balance of your microbiome.

Another common cause of diarrhea is food intolerances and sensitivities. You eat something that disagrees with you—or maybe even violently argues with you—as you try to digest it. Common causes include very spicy foods, the lactose in milk and milk products, and gluten. Some people also have trouble digesting sugar alcohols such as sorbitol, which is found in apple juice and also in many diet candies and chewing gums.

In trying to figure out what's upsetting your stomach, I recommend keeping a food diary. Write down what you eat every day and also any symptoms such as pain, bloating, or diarrhea that occur. If you do this consistently for a week or two, the information gathered can help you and/or your doctor draw some correlations and identify food triggers. A special note on gluten: If you suspect you may have celiac disease, *do not* cut out gluten before you get tested. The tests for celiac disease become inaccurate if you are on a gluten-free diet.

Sometimes diarrhea is caused by parasites such as giardia, which is spread by the feces of infected animals (it's sometimes called beaver fever). It's passed to people when they swallow contaminated drinking water or swallow water from a lake or stream. Another cause to consider: drugs such as antibiotics or antacids that contain magnesium. And let's not forget stress, a frequent cause of the runs.

PASSPORT TO THE TOILET

I definitely have the travel bug, but what I try to avoid at all costs are the other kind of travel bugs! Traveler's diarrhea is the most common travel-related illness—this is one souvenir you'd rather not bring home. You're at highest risk when visiting the developing world, but it can happen anywhere as you expose your digestive system to unfamiliar bacteria and parasites. Although it can definitely wreck your trip, thankfully traveler's diarrhea is rarely life-threatening. To reduce your risk, take precautions. Ideally, eat only cooked foods served hot. Eat raw fruits and vegetables only if they've been washed in clean water or are peeled. Drink water and beverages only from factory-sealed containers. Watch out for ice—it may have been made from contaminated water.

Wash your hands a lot with soap and water when available; use alcohol-based hand sanitizer if you can't wash.

You can still be an adventurous eater and enjoy street food on your travels. Hot foods and drinks prepared in front of you and consumed right away are usually fine. Use your common sense for everything else.

It's possible that in your travels you've managed to pick up a hitchhiker in the form of an intestinal parasite, like the amoeba that causes amoebic dysentery (not fun). If you have traveler's diarrhea that won't go away or if you start getting diarrhea after you return from your trip, see your doctor to get your stool tested. Be sure to mention where you've been.

SELF-HELP FOR DIARRHEA

When you have an attack of acute diarrhea, you lose a lot of fluid quickly. If you lose more fluid than you take in, you can become dehydrated. So, you need to replace the lost fluid and the lost electrolytes (sodium, potassium, magnesium, chloride) as quickly as you can.

If you're able to keep food down, hydrating can be as simple as just drinking a lot of plain water. If you can't eat, or if you want to be sure you're replacing electrolytes along with fluid, try diluted, unsweetened fruit juice, sports drinks, broth, or water mixed with oral rehydration salts (you can get these in any drugstore).

Once you feel ready to eat, I recommend a bland diet. This is a modified form of the popular BRAT diet. The BRAT diet was originally meant for kids (not just spoiled ones), but many adults use it as well. *BRAT* stands for bananas, rice, apples (or applesauce), and toast. The bananas and apples are good for electrolytes, and all the foods are easy to digest. Because there isn't a lot of nutrition in the BRAT diet, I recommend adding in clear broths, salty pretzels and crackers, and potatoes (minus the butter or cream). Eat a small meal every few hours and keep drinking. Foods to avoid while you are having acute diarrhea are raw vegetables, because you want to rest your bowel, and dairy products, because you may be temporarily lactose intolerant. Also lay off the alcohol, caffeine, spicy foods, and sweets until you're feeling better.

You can slow down the diarrhea with pink bismuth (Pepto-Bismol) or the nonprescription drug Imodium (loperamide). However, while they often help stop the diarrhea, they also prevent you from expelling the offending agent. If you have food poisoning and the diarrhea is not very frequent or painful, it's better to let the contaminated food pass out of you quickly.

DIARRHEA AND YOUR MICROBIOME

A bout of diarrhea from a stomach bug or food poisoning can throw your gut microbiome off a bit, but it will usually recover fairly quickly. To help it along, as soon as you feel able, up your intake of prebiotic and probiotic foods for a few weeks or take supplements.

If your diarrhea is found to be caused by a bacteria or parasite that's still sticking around after a few days, your doctor may prescribe an antibiotic. Taking antibiotics for an illness or infection can have a wipe-out effect on your gut microbiome. In addition to killing off the bad bacteria, antibiotics kill off the good ones, sometimes to the point of causing diarrhea—the very thing you're trying to treat. Don't stop taking the antibiotic if prescribed—you need it. Instead, follow the self-help steps to avoid dehydration and keep you feeling as well as you can under the circumstance. Eating plenty of prebiotic and probiotic foods and taking supplements along with the antibiotic can help keep diarrhea to a minimum. Continue with the increased probiotic foods and supplements for at least a week after you're done with the antibiotic.

Burping, Bloating, and Gas, Oh My!

If you think talking about your stool is embarrassing, try talking about gas. Still, talk we must; passing gas is a totally normal bodily function. The average person emits gas from one end or the other at least a dozen times a day.

Think of your digestive tract as a big balloon. When food and gas fill the balloon, it expands, and you feel stuffed or bloated—that "can't button my pants" feeling.

Belching and burping are usually caused by swallowing air in the normal process of eating. You may also get extra air in your stomach from carbonated beverages, from eating quickly, using straws, and from sucking on hard candies or chewing gum. When you feel pressure in your stomach from the trapped air, you burp to release it.

Some of the air you swallow will end up getting passed into your small intestine. From there it goes into your colon and then gets expelled. You generate additional gas in the colon itself when your bacteria ferment fiber and undigested food bits. The process releases hydrogen, carbon dioxide, and methane gases as by-products. Surprisingly, most of the gases produced in your colon get absorbed back into your body through the mucus lining. They pass into the bloodstream and you eventually breathe them out. Only about 20 percent go out as flatulence. Most times, really odorous gas comes from sulfur compounds released by the colon bacteria.

Excess gas is often more embarrassing than anything else. But it can be a sign of lactose intolerance, fructose intolerance, or gluten sensitivity, and it's a common symptom of IBS.

To cut back on burping and gas, cut back on carbonated drinks like soda and beer. Diet candies containing sugar alcohols can also be a gas trigger. Some foods, like beans, are notorious for producing intestinal gas from bacterial fermentation. Cruciferous vegetables like cabbage, broccoli, and cauliflower contain sulfur and contribute to smelly gas. Eggs and meat can also cause that sulfur smell. If you're having more gas than usual, you can try cutting back on your intake of the most common culprits.

Nonprescription products that reduce gas production can often be very helpful. Simethicone (Gas-X and other products) is an antifoaming agent that works by letting the small gas bubbles in your digestive tract join up to form fewer but bigger bubbles, which are easier to pass. Dietary supplements that add enzymes to help break down gas-producing foods before they reach your colon can also work. Lactase supplements (Lactaid) help with lactose intolerance. Supplements with the digestive enzyme alpha-d-galactosidase (Beano) help with breaking down the hard-to-digest carbohydrates in beans, vegetables, and grains. Take these supplements just before you eat these foods.

The metabolic activity of your gut bacteria is responsible for most of your intestinal gas. Some people naturally have more bacteria that produce methane gas, so they tend to have more flatulence. Can prebiotics and probiotics shift the balance away from these bacteria so less gas is created? Very likely. In one study, prebiotics helped about as much as going on a low FODMAPs diet. It's certainly easier and probably better for your long-term nutrition to take a prebiotic supplement than stay on the FODMAPs diet long-term.[3]

A totally different approach to excess gas is to attack it right after it comes out. As a gastroenterologist, I am sent the wackiest things in the mail to try. One product I recently received was a "disposable gas neutralizer," a pad you place in your underwear. Yup, some women get sent flowers; I get butt pads. The pads contain carbon filters and are meant to absorb gas, especially the foul-smelling kind. Other versions contain charcoal, which does the same thing. While there isn't great data yet on these, logically they make sense and are worth a try if you want to stop gas in its tracks.

Irritable Bowel Syndrome

Okay, you've just read all about constipation, diarrhea, burping, gas, and bloating. What if you had a condition that gave you all of them, plus abdominal pain, unpredictably and often? You would have irritable bowel syndrome (IBS), like about 12 percent of the U.S. population.

IBS is a constellation of symptoms, including abdominal pain and changes in bowel movements (either constipation, diarrhea, or both). People suffering from IBS often feel bloated and like they haven't finished moving their bowels; they may also pass a lot of mucus in their stool.

One of the most frustrating aspects of IBS, for my patients (and for me), is that they don't have anything obviously wrong with their colon. All the tests we do come back normal, which means it's still unclear what actually *causes* IBS, and makes it a condition we have to diagnose by excluding everything else that could be causing the symptoms. Many experts think it's caused by a problem with your gut–brain link—your brain and your gut aren't coordinating well. IBS makes your gut more sensitive than usual, so, for example, with the same amount of gas in your intestine, you feel abdominal pain more than someone who doesn't have IBS. Also, when your gut and brain don't work together well, the muscles in your bowel don't always contract the way they should. They can move too slowly, causing constipation, or too quickly, causing diarrhea, or do both at different times.

An alternative explanation for the cause of IBS is changes in the colon microbiome. A number of studies have shown that people with IBS have a different group of predominant bacteria in their colon than people without IBS. It's still unclear, however, if the changes are causing the IBS or are a result of it. Also, some of the studies looked at people who developed IBS after a severe gastrointestinal infection, and the changes that occurred in the microbiome.[4] I see postinfectious IBS often in my practice. A bout of IBS is triggered by an intestinal infection. The infection is gone, but the IBS symptoms can last for months or even years.

Managing IBS

Managing IBS depends, in part, on the type you have. The use of medication to deal with the constipation symptoms of IBS would be different from the medications for diarrhea-predominant IBS, for example.

On the diet end, I recommend eating more fiber and avoiding gluten. Many patients also benefit from following a low FODMAPs diet when they have a flare-up of symptoms. (Check back to chapter 3, "The Kitchen," for more on fiber, gluten, and FODMAPs.) Research suggests that soluble fiber is more helpful for IBS symptoms than insoluble fiber—in fact, insoluble fiber may actually set off IBS symptoms.[5] Eating more oatmeal and fruit can be much more helpful than eating more salad greens, because the soluble fiber in these foods holds water (helping with constipation and diarrhea) and isn't fermented as much by your gut bacteria (helping with gas, bloating, and abdominal pain). Add fiber to your diet *gradually*, just a couple extra grams a day. Any faster than that and you're going to get a lot of gas and bloating, and maybe constipation or diarrhea—and that could trigger your IBS symptoms.

For some reason, people with IBS can be sensitive to gluten from foods that contain wheat, barley, and rye, even though they don't have celiac disease. Avoid breakfast cereal, bread, pasta, and processed food.

Lifestyle changes can also help, perhaps because they improve the communications between your brain and your gut. As much as you can, I recommend being more physically active (see chapter 6, "The Home Gym"), lowering your stress level (difficult, I know, but see chapter 7, "The Zen Corner"), and getting enough sleep (see chapter 8, "The Bedroom").

Before they end up coming to me, many of my patients have tried all sorts of supplements and alternative treatments, such as traditional Chinese herbs, acupuncture, and even reflexology. There is some evidence that hypnotherapy and acupuncture can help,[6] and the evidence for yoga and mindfulness training is strong enough that I recommend them for my patients.[7] In the herbal supplements area, peppermint oil capsules are the one herbal remedy that does seem to help, especially with pain and bloating.[8]

Because of the link between IBS and microbiome changes, one of the first lines of treatment that I employ for IBS is trying to rebalance the microbiome. Probiotics along with prebiotics have been shown to help reduce IBS symptoms. The evidence here is fairly good, but the studies use a lot of different products and bacteria strains, so it's hard to compare them. I recommend both types of supplements as well as probiotic-rich foods to my IBS patients because they often do help.[9]

SIBO

One condition that is thought to cause IBS in some cases is SIBO, or small intestinal bacterial overgrowth. SIBO happens when large numbers of bacteria grow in the small intestine, usually in the area near the ileocecal valve, which connects the small intestine to the colon. Ordinarily, your small intestine has relatively few bacteria in it. It's a barren desert of bacteria compared to the lush rain forest of bacteria in your colon. You don't want any extra bacteria growing in your small intestine, because they can take up nutrients that should be going to the rest of your body. As the bacteria break down nutrients, the by-products can damage your small intestine, leading to leaky gut syndrome. You'll also probably have a lot of bloating, cramping, gas, and diarrhea. You're more likely to develop SIBO after abdominal surgery or if you have a disease like diabetes that slows down the passage of food through your digestive tract, but it can occur even without these risk factors. We can check for SIBO with some special breath tests and treat the overgrowth of bacteria with antibiotics.

Hemorrhoids

So now that you have a handle on your small and large intestines and the gas and stool contained within, let's move on to the actual end of your digestive tract: your anus. And what do we often find at this end? Hemorrhoids. Hemorrhoids are swollen, inflamed veins located under the skin of your anus or in the lower part of your rectum. We all have tiny veins in this area, but if something exerts pressure on them, these veins can swell into hemorrhoids. Hemorrhoids fall into two categories: external and internal. External hemorrhoids form under the skin around your anus. They can cause anal itching, a lump near your anus, and anal pain that's worse when you're sitting down. Internal hemorrhoids form in the lining of your rectum. You usually notice them when you see small amounts of bright red blood on the toilet paper, in your stool, or in the toilet bowl after you have a bowel movement. Internal hemorrhoids can prolapse, or fall through the anal opening. *That's* when they hurt.

Can you have both types? Yes—most of my patients do. Can you have them and not realize it? Absolutely. Listen to what happens to me almost every day. I perform a colonoscopy on a sleeping patient. When they wake up, I tell them the good news that everything looked good in their colon. All I saw was just some small hemorrhoids. Cue the disbelief. Hemorrhoids? Me? No way!

Hemorrhoids have a lot of causes. The main cause is straining to have a bowel movement, which puts a lot of pressure on the veins. Similarly, chronic constipation or diarrhea, which leave you sitting on the toilet for long periods, can cause hemorrhoids.

A low-fiber diet that makes your stools harder to pass can also be to blame. And if you're pregnant, the extra pressure of the baby, as well as increased blood flow, can cause the rectal veins to swell, leading to hemorrhoids. Hence the lovely donut pillow that women are sent home with from the maternity ward—not only is the vagina sore the first couple of weeks as a new mom, but painful hemorrhoids can make sitting excruciating as well. The good news is that with time as well as self-care measures, the pain from pregnancy-induced hemorrhoids subsides.

Sometimes hemorrhoids get so annoying and painful that they have to be surgically removed, but most people can avoid that extreme. You can easily manage hemorrhoids by adding fiber to your diet to make bowel movements faster and easier. Itching and pain can be managed with nonprescription ointments that contain witch hazel or a mild corticosteroid. If the symptoms are painful, warm baths or sitz baths (a soaking apparatus that attaches to your toilet) several times a day can really help. But if the hemorrhoids aren't bothering you—they're simply there—then I suggest you forget about them and leave them alone. Save your energy for other parts of your Gut Renovation.

GAME OF THRONES

You really should only be sitting on the toilet for as long as you have the urge to defecate, usually ten to fifteen minutes. Loitering on the porcelain throne while you're reading or scrolling through your phone (unsanitary!) can put increased pressure on your rectum, which can cause hemorrhoids. If you thought you had the urge but it's not there when you sit on the toilet, don't push and strain anyway or wait too long for the urge to come back. This is not the time or place to catch up on your emails or your magazine reading or play a video game. When it comes to the bathroom, just do your business, wash your hands, and get out.

Diverticular Disease

Holes in your pipes? We call this diverticulosis. You probably don't know it, but you may already have diverticula in your colon. These are little pouches that make your colon lining look like Swiss cheese, although the holes don't penetrate the lining, thank goodness. These pockets push outward in weak areas of the colon wall, most commonly in the lowest portion. About 35 percent of people under age sixty already have them; nearly 60 percent of people over age sixty have them. The outpouchings are generally only about the size of a pea and don't usually cause any problems. Sometimes, however, bacteria or fecal matter can get trapped in the pouch, causing inflammation and infection—you have diverticulitis. If this happens you may have abdominal pain, fever, constipation or diarrhea, perhaps with nausea and vomiting. You may even have rectal bleeding. The symptoms will probably be scary enough to send you to the emergency room—and they should. In rare cases, diverticulitis requires urgent surgery, so don't sit at home with this. More common, the treatment is antibiotics, rest, and a liquid diet for a few days. In more severe cases, you may need to spend a few days in the hospital with IV antibiotics.

This is another gastrointestinal disease where your Gut Renovation will play a positive role. A high-fiber diet may help keep problems from developing if you already have diverticulosis. A healthy gut microbiome may also help, possibly by keeping down the bad bacteria that might find a home inside a diverticular pouch. If you have the pouches of diverticulosis, probiotic supplements may help prevent diverticulitis.[10]

For years, we would tell people with diverticulosis to avoid eating nuts, seeds, and fruits with tiny seeds like raspberries. The thinking was that undigested particles would find their way into the outpouchings and cause infection. This turns out not to be true, so keep eating those delicious and nutritious high-fiber foods.

Colonoscopy and Colon Cancer Screening

Colonoscopy, my favorite subject! Maybe not so much for you, but a colonoscopy could save your life when used as a screening tool. Just as your home needs some routine maintenance and checks to keep a small problem from blooming into a big one, so does your colon.

A colonoscopy is a procedure that uses a long, flexible tube with a light and tiny camera at one end to look inside your rectum and colon. In colon cancer screening, we're looking for tumors or polyps that could become tumors. Most colon cancers start as polyps, which is actually a great thing. It means that if we find and remove a polyp from your colon, we can stop colon cancer before it starts. Of course, while we're in there, we look for anything else that might become a problem, like diverticula or inflammation.

Your bowel has to be empty before we can do a colonoscopy so we can see properly. The clean-out process starts the day before the procedure. It's not so fun, because it involves skipping dinner and then drinking down a liquid laxative solution. You then get to spend some quality time with your toilet emptying out your entire colon. Early the next morning, tired and hungry, you head to the gastroenterology clinic to see a gal like me. We get you into a stylish gown and slippers, insert an IV in your arm, and then put you to sleep with a short-acting anesthetic that will keep you in La La Land for the duration of the procedure. For most people, the actual colonoscopy takes only about fifteen to thirty minutes. After the test, you have to stay in the recovery area for about an hour. During that time, you'll pass gas, maybe very audibly. We can't let you leave until we know you have, so don't hold back. You'll also be given some snacks (if the doctors haven't eaten them all), which you will really enjoy because by now you're starving.

Why are you putting yourself through this? Because colon cancer is the second-leading cause of cancer death in the United States—and most cases are very treatable if caught early enough. Also, rates of colon cancer in people under fifty are on the rise, so getting screened early is a very good idea. A recent study suggests that an important risk factor for colon cancer in this age group is a history of exposure to antibiotics.[11] That's almost everyone, so people, please get screened!

Some people avoid getting a colonoscopy because they have no digestive complaints. They think that since they feel good, they must not have colon cancer. But colon cancer often doesn't cause symptoms until it's quite advanced. Polyps almost never cause any symptoms, which is why we screen everyone at the appropriate age, regardless of how they feel.

Current guidelines from the American Cancer Society say that if you're at average risk for colon cancer, you should get screened starting at age forty-five and then once every ten years through age seventy-five. If you're at higher risk, you may need to be screened more often, based on discussions with your doctors. You're at higher risk if you have a strong family or personal history of polyps or colon cancer, if you have inflammatory bowel disease, or a genetic colon cancer syndrome. To clarify, these screening guidelines are for people without symptoms. If you do have any symptoms of colon cancer—rectal bleeding, new onset of abdominal pain, constipation, diarrhea, weight loss, narrow or pencil-thin stools—then you should see a doctor ASAP.

The yuck factor alone shouldn't keep you from having a colonoscopy—this procedure remains the best screening tool for colon cancer. But for a range of reasons, like too much risk from the anesthesia, some people really can't have one. If that's the case, an alternative is a virtual colonoscopy using CT scans (you still have to do the bowel prep). You can also talk to your doctor about stool tests that look for blood and genetic markers of colon cancer.

There are other ways to renovate your gut to reduce your risk of colon cancer, but those should really be used in addition to regular colonoscopies, not as a replacement. Don't smoke, drink alcohol only occasionally, and get regular exercise. Limit your intake of red meat, skip processed meats altogether, and eat a diet with plenty of vegetables, fruits, whole grains, calcium-rich foods like beans, and yogurt. We already discussed the link of dysbiosis and cancer, so keeping that microbiome balanced is also important.[12] The Gut Reno eating plan at the back of this book has you covered on all fronts.

Stool Testing

Until DNA analysis became as cheap and easy as it is today, figuring out exactly what types of bacteria you had in your gut was very challenging. Almost all your gut bacteria are anaerobic, meaning they don't thrive in an oxygen-containing environment. That makes them very hard to grow in a lab. But DNA testing allows us to identify bacteria without having to culture them, which is why our understanding of the gut microbiome has been improving very rapidly in recent years.

The same technology lets you send a sample of your own stool to a lab for analysis. A number of commercial companies have sprung up to offer this service. You send your sample to them in the mail and get back a report about which bacteria are in it and in what ratios compared to other people. Is this useful information? Hard to say. It's a snapshot of your gut bacteria at one particular moment. Your microbiome varies considerably from day to day and even hour to hour, depending on what you've been eating and doing. How it looks right now doesn't necessarily tell you how it looks over time.

There are some valid medical uses for stool testing, like looking for parasites and bad infectious bacteria, but unless your doctor recommends it, commercial stool testing is mostly experimental at this point. Do I see a future where we are able to make informed decisions based on stool microbiome analysis? Yes, but I don't think we're there quite yet.

Bathroom Renovations

Real bathroom renovations can be very helpful for fixing and even preventing a lot of gut problems. Some are the simple, inexpensive kind you can easily do yourself. Others may involve some expense and a good plumber, but they'll pay off in better gut health. In my own home, we have the products described below. I've road-tested them and they actually work.

POTTY TRAINING

So you think you learned how to use the toilet properly as a toddler? Think again. Now, before you start cursing your mother (honestly, we moms get blamed for everything!), let me explain.

How we sit when we defecate can affect how easily the stool comes out. The ideal position is actually to squat. This position opens up your hips, which then allows the rectoanal canal to straighten. Voila! That stool just plops right out. But of course toilet design doesn't allow for easy squatting, which is why you were taught to sit on it. Your mom didn't want you perched on top in a squat, potentially falling in! But this predicament is why some smart entrepreneurs (gotta love the American hustle) invented a really simple Gut Renovation tool: a defecation posture modification device, better known as a toilet stool. You place the stool (it's usually seven or nine inches high) at the base of the toilet. When you sit to have a bowel movement, the stool raises your knees and puts you into more of a semisquatting position. The position helps reduce straining and lets you empty your bowels quicker and more completely.[13] I often recommend this stool for patients who have constipation, painful bowel movements, and/or hemorrhoids.

The next renovation can range from easy and inexpensive to complicated and very expensive: a bidet. This bathroom fixture is very common in Europe and Asia and is starting to become a lot more popular in the United States. A bidet is a low wash basin located next to the toilet. You sit on it to clean yourself after you use the toilet. Like any wash basin, it has hot and cold running water, but the stream comes from below or the side, not above.

During my childhood I spent my summers at my grandmother's house in Sri Lanka, which had bidets in every bathroom. Although it felt very strange the first few times I used it, I soon became a bidet addict. Today, I like bidets because they help you keep clean gently, which is great in general but really great if your bottom is sore from hemorrhoids or frequent bowel movements. I also recommend bidets for my hyper wipers: patients who overuse toilet paper because they never feel quite clean enough and irritate the skin around their anus in the process.

Adding a bidet can be expensive, and not every bathroom has space to install one. That's why I love the newer bidet attachments that you can use with your own toilet. These range from very simple versions you can install yourself to elaborate attachments that need an experienced plumber.

GUT TOOLBOX 

- *Make sure it's brown before you flush it down.* I encourage you to visually inspect your stools before you send them away into oblivion. Note the color (red and black being the most worrisome) and pay attention to the size, shape, consistency, and smell.

- *Talk it out.* Bathroom issues such as bowel movements and gas are embarrassing to talk about, but everybody has them. Don't hesitate to bring them up with your doctor.

- *Lube your tube.* If you suffer from constipation, increased hydration can help things move along more smoothly. Set an alarm to go off at least four times a day and remind you to drink one to two extra glasses of plain water. And eat more fiber!

- *Ditch the dairy if you're battling a bout of diarrhea.* Avoid dairy products until your normal bowel movements resume.

- *Track your food.* To deal with food intolerances or sensitivities, use a food diary app like YouAte to easily keep track of what you eat. Once you identify a culprit, cut back on it or avoid it. Foods to watch for include nuts, eggs, milk and dairy, high-fructose foods, and foods on the FODMAPs list.

- *Friends with benefits.* Make a pact with similar-aged friends to screen for colon cancer with a colonoscopy. This is a must-do—seriously, it could save your life. Once you've all completed your colonoscopies, treat yourselves to a fun night out together.

CHAPTER 5

THE POWDER ROOM: BEAUTY ISN'T JUST SKIN DEEP

By now, you understand the role your gut plays in maintaining the optimal functioning of virtually all your organ systems. If the bathroom is where the serious inner plumbing occurs, the powder room is where our serious vanity is indulged. You may literally powder your nose here or maybe you just stare in the mirror, playing that fun game where you count the number of dark spots or wrinkles that have appeared since the last scrutiny. Regardless, in your Gut Renovation, the powder room is your skin. Not only is the skin—the largest organ in your body—an important barrier to toxins and other assaults on your body, but it's also a reflection of your inner health, and the first thing outsiders look at when assessing your age and vitality.

What goes on with your skin is by no means just superficial. As it turns out, your gut and your skin are tightly linked. In fact, because what happens in your gut has such an impact on your skin, we call that interplay the gut–skin axis. Like your gut, your skin has its own microbiome—you're literally covered with bacteria, from your scalp to your toes. Because of the gut–skin axis, what happens in one microbiome affects the other. Both your gut and your skin are links between the external world and the internal body; both have rich blood supplies; and both communicate actively with the nervous system, the immune system, and your hormones.

In my practice I often prescribe probiotic supplements to my patients to help them recover from digestive problems that are caused by an imbalance in their gut microbiome. After they've been taking them for a while, they almost always mention to me that their skin is looking and feeling better. Acne, rashes, itchy spots, dry skin—they're much improved or even gone. Even their hair and nails look better. They're often pleasantly surprised by this. I was surprised too when I first observed this phenomenon years ago, but now I know better.

To understand how the gut–skin axis works, let's take a closer look at your skin.

Your skin is way more than pretty wrapping paper for your body. It's a barrier to protect you from the outside world. It cushions your body and holds in your body fluids, regulates your body temperature, protects you from disease, and eliminates waste. It's densely packed with nerve endings that let you feel and respond to the world around you. Not only is your skin your largest organ—it's the only one you can actually see.

But what you see of your skin is only its outermost part, the layer known as the epidermis. The cells of the epidermis are made mostly of keratin, a fibrous protein that's also the main material for your hair and nails. The epidermis gives your skin its strength and ability to hold in moisture—these are the cells that absorb water and make your fingers look wrinkled after a bath. A key role of the epidermis is to act as a barrier that keeps the bacteria on the surface of your skin from entering deeper into your body. The cells of the epidermis are constantly being shed and replaced by new cells pushing up from below. Because epidermis cells are just at the surface of your skin, they're not connected to your circulatory system (your bloodstream).

The outer layer of your skin rests on a lower layer called the dermis. This supporting layer is sometimes called the true skin because it's full of blood vessels and nerve endings—the structures that connect it to the rest of your body. The dermis is also where your hair follicles, sweat glands, and sebum (skin oil) glands are found. Crucially, the dermis is where collagen, the protein that gives your skin firmness, and elastin, the protein that gives your skin elasticity, are most abundant. The dermis also contains lots of glycosaminoglycans, or GAGs, complex proteins that support collagen and elastin and help keep moisture in your skin. You can think of these proteins as the scaffolding that supports firm, healthy skin. Because collagen and elastin breakdown are a part of the wrinkle-forming process, you can see why this layer of skin plays a big role in how well (or not) your skin ages. (More later on this—I promise!)

Aging and Your Skin

As your skin ages, it inevitably becomes duller, sags, wrinkles, and gets age spots and dry patches, as well as under-eye bags. Or does it?

I'm not going to make false promises here. Yes, as you age, your skin does change, and not for the better (although at least your acne probably improves or goes away). That's intrinsic aging, the sort you can't really do much about. It's just a medical fact, for example, that after your early twenties your body produces about 1 percent less collagen every year. Your oil and sweat glands don't work as well, you make less elastin, you make fewer GAGs, and you lose supporting fat from the dermis layer.

But the skin changes from intrinsic aging are mild—a few laugh lines and a greater need for moisturizer are usually the biggest issues.

The real damage to your skin—the real source of wrinkles and every other appearance of aging on your skin—is extrinsic aging, which is aging from factors outside your body. Take a moment to let that sink in. How you age cosmetically has more to do with factors within your control than with the number of candles on your birthday cake—it's all about UV rays from sunlight, air pollution, smoking, alcohol, diet, stress (see chapter 7), and lack of sleep (check chapter 8). But luckily, you have control over your exposure to many of these skin saboteurs.

When it comes to diet, there are two key factors here: load up on vitamin C and keep sugar to a minimum. Vitamin C is essential for building collagen and keeping it strong. Include plenty of vitamin C in your diet and you support your collagen production. More collagen, fewer wrinkles. Get vitamin C from a diet rich in fruits and vegetables. Sugar is a little more complicated. When your diet is high in sugar, the excess floats around in your bloodstream and ends up glycating (sticking to) proteins. You basically caramelize your body with new molecules called advanced glycation end products, or AGEs. When AGEs form, they can damage other proteins like collagen. Specifically, they make collagen weak, brittle, and less elastic, so it no longer supports your skin as well. Wrinkles follow.[1]

Of course, collagen is a major component of healthy, youthful skin. Is there a way to increase your supply? For years, there were questions about whether collagen is actually absorbed into your body if you drink or eat it. And even if it is absorbed, does it then actually reach your skin? Turns out recent research has answered yes to both of these questions. Hydrolyzed collagen (a more easily absorbed form) in supplements, powders, or beverages has been shown to increase collagen production in the skin, which we know can have a major effect on reversing aging changes. Fine lines and wrinkles are reduced, and elasticity and hydration are increased. Cheers to younger skin![2]

Inner Health, Outer Glow

When my patients start to restore a healthy gut microbiome, those improvements are often visible to me. From visit to visit, I see complexions improve, hair get thicker, or rashes fade away.

So if fixing an unbalanced gut microbiome and becoming healthier inside helps disease-related skin problems so much, what could it do for counteracting skin agers like UV light, dryness, and even stress? What about acne, eczema, even dandruff? As it turns out—a lot.

The gut microbiome affects your skin directly through its complex role in your immune system. When your gut is functioning normally, your immune cells function normally as well, keeping down inflammation throughout your body. But when your gut isn't working well the barrier function breaks down. Bacteria and bacteria by-products that should stay safely within your small intestine and colon escape into your circulation and can end up triggering inflammation in the skin. When that happens, the skin's normal functions get disrupted. You could get rashes, pimples, itchy spots, dry areas, or even the rough, scaly skin of psoriasis.

Problems in your gut can also throw off your skin's pH, or acidity/alkalinity balance. Your skin normally is slightly acidic, which helps keep bacteria out and moisture in. As we age, our skin's pH tends to rise out of the optimal range to a higher, more alkaline one. When it's too alkaline, your skin gets red and flaky. The reverse is also a problem—when your skin becomes too acidic, it encourages inflammatory skin conditions, like eczema and acne. There are also issues like allergies; when you're exposed to allergens, your skin might quickly erupt in itchy hives or a rash. Less obvious, because it's slower to appear, is the link between what you eat and skin problems such as acne and rosacea.

But when your gut microbiome is happy and healthy, your immune system is calm. Your overall inflammation level is low—and when inflammation is low, your skin benefits. It gets thicker, stays hydrated better, and is less sensitive, plus your hair gets thicker and glossier. You get that outer glow from your inner health.[3]

The gut microbiome and your skin microbiome "talk" to each other, probably a lot more than we realize right now. To take one example: Your gut microbiome digests fiber from your diet. As a byproduct of digestion, the bacteria produce short-chain fatty acids (SCFA), including one called propionate. It turns out that propionate from the gut reaches the skin microbiome, where it's really good at killing staph bacteria that cause serious antibiotic-resistant infections. So, if your gut bacteria are happy because you feed them lots of fiber from a healthy diet, you're protecting yourself against a possibly deadly bacterial infection that originates on your skin.

Your gut talks to your skin—and your skin talks back—through several mechanisms.

First, your gut absorbs nutrients that can have a direct effect on your skin. If you take vitamin E supplements, for example, they get delivered to your skin through the sebaceous glands. If your nutrition is poor, it shows in your nails and hair, which can become brittle and dry. Going in the other direction (your skin to your gut), if you use a contraceptive patch, pain relief cream, or some other drug on your skin, it's absorbed into your body and will eventually pass out of you through your gut. The same is true for the chemicals in any makeup, cream, or other product you put on your skin. (More on cosmetics and skin care products later on in this chapter.)

Another mechanism for the gut–skin axis is food that shifts your hormonal balance. Among other impacts, this is why eating junk food can make your acne flare. Unlike the commonly held belief that eating oily foods causes acne, it's actually much more likely to be triggered by food loaded with processed carbohydrates and sugar. Those make you produce more of a hormone called insulin-like growth factor 1 (IGF-1), which circulates in your bloodstream. When IGF-1 reaches the sebaceous glands in your skin, it triggers them to produce more oil, which in turn triggers your acne.

The link between your gut microbiome and your skin also plays out through your intestinal barrier. If your gut microbiome is less than optimal, it's easier for substances in the gut to cross the intestinal lining and enter your circulation, thereby promoting inflammation throughout the body, including the skin. This may be an underlying cause of severe acne, rosacea, and even psoriasis. It may also be an underlying cause of the skin problems that go along with some kinds of digestive diseases, like the skin lesions that can accompany Crohn's disease, the distinctive rash that goes along with celiac disease, or the clear link between eczema and food allergies.

Skin-vincible

I've already explained the basics of eating gut-healthy foods and the importance of staying hydrated (check back to chapter 3). When it comes to your skin, what's good for your gut microbiome is also good for your skin and your skin microbiome. Beyond the basics and the prebiotics and probiotics you should now know and love, some foods are particularly helpful to keep your skin soft, clear, and wrinkle-free.

Your skin protects you against the outside world, but in the process it's exposed to ultraviolet (UV) rays in sunlight, air pollution, moisture-sucking dry air, bad bacteria, and other damaging attacks. To fight back, your skin has a lot of protective tools. Antioxidants made in the dermis and also carried to the skin from the circulation fight the damaging free radicals created by UV rays and pollution. Your sebum glands produce skin oil to keep your epidermis soft and moist; epidermal cells are also great at absorbing water from your body. And a healthy, balanced skin microbiome contains so many beneficial or neutral bacteria that the bad ones have trouble multiplying to the point of being harmful.

Your skin can't do it all on its own—it needs you to eat the right foods to support it. What are they? Foods that are high in natural antioxidants such as lycopene and resveratrol, beta-carotene (vitamin A), and vitamin C. Your skin needs plenty of antioxidants to counteract UV radiation and other skin stressors. While your body makes antioxidants naturally, at times you may need more than your body can easily produce. And the older you get, the more you need, both to counter the cumulative damage to your skin and to make up for a naturally reduced ability to make antioxidants. That's where all those vegetables and fruits come in—they're crammed with natural antioxidants, because they grow in the sun and need to protect themselves from UV radiation, just as you do.

All fruits and veggies contain a range of natural antioxidants, but some contain specific antioxidant compounds that we know help your skin. Red or pink fruits and vegetables contain lycopene, a powerful antioxidant that's great for helping to protect your skin against sunburn and UV damage (to say nothing of its potential to protect against heart disease and cancer). Tomatoes, watermelon, pink grapefruit, mangos, guavas, papaya, and sweet red peppers are all also high in lycopene.[4]

Resveratrol is another natural plant-based antioxidant. It's found in red grapes, red wine, peanuts, cocoa, and berries, including blueberries and cranberries. In addition to being an antioxidant, resveratrol has a superpower: it can activate a gene called SIRT1 in human skin cells that helps fight aging by enhancing your skin's ability to resist DNA damage and repair itself.[5]

Vitamin C, found in most fruits and vegetables, is a major antioxidant and plays an important role in preventing photoaging. In fact, when your skin is exposed to sunlight, it almost immediately starts using up the vitamin C in its cells as a way to defend you against the UV rays. To maintain the photoaging protection of vitamin C, you need to keep replenishing it through your diet.

To get even better natural protection against photoaging, make sure you get plenty of vitamin E in your diet. Research shows that vitamin C and vitamin E work together to increase sun protection. Foods high in vitamin E include nuts (especially walnuts), wheat germ, seeds (sunflower seeds, for example) and green leafy vegetables.[6]

Beta-carotene, an antioxidant found in orange foods like carrots and sweet potatoes, is converted in your body into retinol, also known as vitamin A. The process happens in the wall of your small intestine. This is another good example of how your gut and your skin are connected, because retinol essentially moisturizes your skin from the inside out. It triggers cells in the dermis and epidermis to produce collagen and elastin, the skin proteins that help your skin keep moisture in and stay supple and unwrinkled. At the same time, vitamin A helps prevent photoaging by protecting collagen from UV damage.

Your skin loves getting vitamin A from your diet, and it also happily absorbs it when it's applied topically as retinoids (different forms of vitamin A). On the skin, retinoids work by making the cells in the outer layer of the epidermis turn over faster, making room for the new cells migrating up from underneath. They also help keep collagen from breaking down. Both prescription and nonprescription skin products with retinoids are used to moisturize and help treat fine facial wrinkles, age spots, and rough patches from photodamage. In prescription form, retinoids are used to treat acne.

Caution: Retinoid skin products work really well, but they can cause dry skin, itching, and irritation. They can also cause birth defects and should never be used if you are or might be pregnant or if you're breastfeeding.

Your skin really loves when you get plenty of omega-3 fatty acids from fish, nuts, and leafy greens. They help improve your skin's barrier function, keeping moisture in and irritants like air pollution out. They also protect your skin from harmful UV rays and reduce the inflammation that's an underlying cause of acne.

Don't eat: processed food, fast food, dairy products. You already know why junk food is so bad for you, but why the ban on dairy products? Dairy foods can have a similar impact on your hormones as sugary junk food. The end result is more oil from your sebum glands, more clogged pores, and more acne blemishes.

One final idea for keeping your skin clear and supple: drink green tea. Polyphenol compounds called epigallocatechin-3-gallate, or EGCG for short, have antioxidant power and block the damage caused by UV rays from sunlight. Drinking a cup or two of green tea every day can help you stay hydrated, slow skin aging, and prevent photodamage.

Skin Conditions and Your Gut

We've been talking so far mostly how your microbiome and your diet impact skin that's basically healthy. But we're learning more and more about how important the gut microbiome is for problem skin as well.

Take rosacea, for example. This common skin condition makes your face red, sometimes with tiny, visible blood vessels and small red bumps that look like acne. Frustratingly, rosacea tends to flare up and annoy you for weeks on end, then go down for a while, and then flare up again. The causes of rosacea still aren't fully understood, but we do know that dysbiosis and SIBO can play a role. With probiotics and a healthier diet, many people see their rosacea reduced.

If your gut is unhealthy, this also means that the bacteria that cause acne, eczema, and psoriasis are allowed to thrive. Here too, a better diet and probiotics can restore the gut balance and ultimately the skin balance as well, helping you clear up itchy rashes, dry skin, and rough patches!

Defeating Dandruff

Your hair, hair follicles, and scalp have their own microbiome. The bacteria, fungi, and yeast that live there generally live in harmony with other, but sometimes the balance gets disturbed. The result can be the itchy scalp and unwanted skin flakes of dandruff. We have all sorts of shampoos that can keep dandruff under control, usually by killing off some of the microbiome, but they don't always work well, and often contain strong chemicals like coal tar or selenium sulfide or zinc pyrithione. What if, in addition to attacking the bad bacteria, you encourage the growth of good bacteria? They'd suppress the bad bacteria, yeast, and fungi and help make your dandruff go away.

That's where probiotics come in. Balancing the bacteria in your gut has powerful positive effects on your skin—and your scalp is really just the skin covering the top of your head. Probiotics help dandruff by helping to improve the skin barrier function, keeping moisture in, reducing itchiness, and making the hair follicles less hospitable to bad bacteria and fungi.[7]

ARMPITS AND BACTERIA

Smelly armpits are an embarrassing problem that comes directly from your skin's microbiome. Your armpits are warm and moist—an ideal place for bacteria to thrive. And they do. In fact, your armpits have their very own microbiome, one that produces body odor. Some people have armpit odor that's quite strong and won't go away no matter what they do. It's possible these people have an imbalance in their armpit microbiome. The balance is tilted toward the bacteria that produce the smelliest compounds. Interestingly, the aluminum in many commercial antiperspirants may actually promote a "smellier" armpit microbiome. Until recently, if you fell into this unlucky category of severe BO, you were doomed to frequent pit washing and lots of clothing changes. Our increased understanding of the skin and gut microbiome, however, gives us some new ways to deal with bad body odor. The low-quality fats in fast foods and processed foods change the lipid balance of your sebum, the oily, waxy secretion from the many sebaceous glands in the armpit. When you eat too much of the bad fats and not enough of the good ones, your sebum becomes more hospitable to the smelly bacteria. Moving your diet in a healthier direction with a better fat balance encourages the growth of bacteria that don't produce bad-smelling by-products.

In severe cases of BO, however, dietary changes alone may not help enough. An armpit bacteria transplant may be needed. Yes, you read that correctly. The idea is to introduce better-smelling bacteria (via sweat transfer) from volunteer donors into the recipients' armpits. So far, the approach is experimental, but the results are promising. In one study, improvements in odor were noted within a month in sixteen out of eighteen volunteers—and the improvement lasted for several months. It's possible that in the future, bacteria transplants for bad BO will become routine. Or maybe we'll just roll on natural probiotics instead of chemical deodorants.[8]

Disrupting the Skin Microbiome

Modern life can play havoc with your skin microbiome. We spray insecticide on our skin to keep bugs off, we expose it to pool chemicals, we put on makeup—all things that can potentially kill off skin bacteria, good and bad, and interfere with the protection our skin barrier gives us. In general, the skin microbiome manages to hang in there and deal with what hits it, but nowadays we're washing our hands a lot more. Soap, cleansers, and hand sanitizer wash away and kill bad germs, but also sometimes disrupt the natural balance of your skin microbiome. All that hand washing—while important—is causing little cracks to emerge on the skin and is giving bad bacteria an entry point for infection. And those little cracks sting when you use alcohol-based hand sanitizer!

To reduce the irritation from frequent hand washing, use mild, fragrance-free soap and wash in warm, not hot, water. If possible, apply moisturizing cream after drying your hands or using hand sanitizer. If you can't use hand cream after every wash, at least slather it on when you can and be sure to moisturize your hands thoroughly at night before bed.

Then there's a new skin problem that may be with us for a while: Maskne, skin that breaks out underneath a face mask. Maskne seems to be caused by the way face masks trap moisture against the skin, especially around the mouth and nose. Moisture-loving bacteria from the skin microbiome take advantage of the situation to get into the hair follicles and sebum glands and give you pimples and irritation. It's the same problem athletes have under tight-fitting gear, like the chin strap of a bike helmet.

To prevent maskne, be sure to wash reusable masks thoroughly and change them if they get wet or sweaty. Wash your face with the gentlest fragrance-free cleanser you can find. Don't wear makeup under the mask.

Probiotics and Your Skin: The TULA Story

What's good for your body is good for your skin. So if probiotics in the gut are good for your skin, what if you used them directly on your skin instead? That's the question I asked myself back in 2014, as interest in the skin microbiome was becoming a movement. The answer was TULA, a skin care company I created to develop innovative products using topical probiotic extracts. Based on cutting-edge research, our products combine extracts of multiple specific strains of beneficial bacteria with superfoods like blueberries and turmeric to help your skin look its best.

One of our most important goals at TULA is to design products that help you avoid and repair photoaging—the age spots, fine wrinkles, and dry areas caused by exposure to UV light. We use several types of probiotic extracts with clinical backing behind them. These extracts have been shown to help improve skin acidity and antioxidant activity and prevent skin fragility and age spots. They help reduce signs of inflammation on the skin, such as redness, blotchiness, and swelling.

And by keeping the skin hydrated and thicker, they help prevent fine lines and wrinkles.

Topical probiotic extracts also help strengthen the skin barrier to keep moisture in and bad bacteria out. When the skin's barrier function improves, so does acne, rosacea, and eczema. The bacteria that cause or worsen these conditions can't get in as easily. Because topical probiotic extracts improve the barrier function, they also help your skin stay moist and flexible—even when you need to use irritating drying lotions to treat acne and other skin conditions.[9] So when thinking about protecting your skin, it's important to incorporate skin care products that are made from probiotics into your routine, or at least use products that aren't harmful to your skin's natural microbiome. The term "microbiome friendly" is a new term in the beauty industry to address this issue—look for it on product labels.

GUT TOOLBOX 

- *Don't skimp on the sunscreen.* Most people don't use enough when they apply it. You should be using at least one shot glass full of sunscreen (with an SPF of 30+), and don't overlook spots like your ears, your hands, your neck, and your chest. Apply every two hours, and use sunscreen 365 days a year—UV rays can penetrate clouds! And don't forget your wide-brimmed hats and sunnies. Not only do they look stylish, but they also help prevent crow's feet by protecting the delicate skin around your eyes.

- *Fight pollution.* While you probably can't move your home out of a polluted area, you can try to avoid walking in high-traffic areas and make sure to thoroughly cleanse your face when you return from a long period outdoors so those environmental toxins don't linger on your skin.[10]

- *Keep your coverage clean.* Makeup brushes and sponges can be a perfect breeding ground for bacteria—and not the good kind. So clean these weekly with warm water and a gentle cleanser. And don't share them with other people.

- *Toss the turtleneck.* When it comes to skin care, many women focus only on their faces and forget about their necks. Your neck skin gets a lot of wear and tear (especially from looking down at devices), and as a result this is a prime area of wrinkles and sagging skin. Make sure you don't stop your skin care regimen at your chin. Keep applying downward and give your neck some love.

- *Don't get skintense.* Stress can bring on hormone changes that trigger breakouts, so when you're feeling under the gun, take a break and practice deep breathing. If you prevent a blemish, that's one less thing to stress about!

- *Beauty rest.* Better sleep leads to better skin, especially when it comes to dark under-eye circles. So make sure to prioritize your zzz's and use a humidifier at night to prevent overnight drying. Dry skin is more fragile, duller, and shows more visible wrinkles.

CHAPTER 6

THE HOME GYM: WORK OUT TO TURN BACK THE CLOCK

Not every home renovation has the space and budget for a home gym. No problem—and also, no excuse. You don't need an elaborate setup to work out at home, stay fit, and age better. In fact, everything I'm going to talk about in this chapter can be accomplished with nothing more than a pair of comfy running shoes, a two-by-six roll-up foam mat, a pair of light hand weights, and a couple of leg resistance bands.

Why Exercise?

I'm a gastroenterologist, but I talk with my patients about exercise almost as much as I do about diet or medication. Why? Because it's good for your gut in lots of ways, including creating positive changes in your microbiome. It's also great for every aspect of the rest of your health.

Current guidelines from just about every medical organization recommend getting at least 210 minutes of exercise over the course of a week. Allow me to do the math for you—that's just thirty minutes a day. Of course, even half an hour of exercise every day can sometimes seem like too much to squeeze in. I understand, because I don't always manage thirty minutes of solid exercise either. When I'm having one of those days, I aim for three ten-minute exercise sessions, worked in when I can. Because let's be real—you can *always* squeeze in ten minutes, here and there. In fact, short bursts of exercise can be just as good—and in some ways better—than a longer workout.

Just as important: Any sort of regular physical exercise helps keep your brain sharp and your mood positive. Exercise is a key step you can take now to help avoid cognitive impairment in later life.[1] As you age, exercise can also help you avoid chronic ailments such as diabetes, heart disease, and many types of cancer. And not surprisingly, at any age, aerobic exercise slashes your risk of death. How long you've been inactive before you begin exercising doesn't matter. Get moving and chances are you'll live not only better but longer.[2] Want a recent real-world example? A study looking at Californians infected with COVID found that those who were inactive were more than twice as likely to be hospitalized or to die from the disease than those who exercised regularly.[3]

Exercise for the Gut

Your gut benefits from exercise in two ways. First, when you're sluggish, so is your digestive system. If you don't get enough exercise, your digestion may slow down, causing bloating, gas, and constipation. Exercise can help stimulate your colon, *and* help you relieve stress (which can make most digestive problems worse).

The other benefit of exercise for your gut is how it changes your microbiome for the better. People who exercise regularly have greater diversity among their gut bacteria—no matter what they eat. The improved diversity has another benefit: When you're fitter and your microbiome is more diverse, you have more of the types of bacteria that produce butyrate. As I explained back in chapter 3, butyrate is a short-chain fatty acid (SCFA) that's the main energy source for the cells that line your gut and maintain its barrier function. High levels of butyrate production are a good indicator of gut health.

It's especially important to take an active role in diversifying your microbiome, but the diversity of your gut bacteria naturally tends to drop with age—and many older adults regularly take medications that may also impact their microbiome. The lack of diversity and fewer butyrate-producing bacteria may be behind some of the reduced immunity and increased inflammation we see in the elderly. Here too, exercise can help. In addition to all its other benefits, research shows how exercise can improve the makeup of the gut microbiome and really make a difference for healthy aging.[4]

Serious athletes have always been interested in diet and supplements as ways to improve performance, so there's a lot of interest in the latest research looking at how exercise affects the microbiome, and vice versa. Whether the studies are of lab rats or "gym rats," what's becoming clear is that if you're a serious sportsperson, your gut microbiome can give you a competitive edge.[5] Studies of elite athletes have shown that they have more gut bacteria that are good at eating lactate, a waste product produced by your muscles when you exercise (it's why you feel the burn when you work out). Because their gut bacteria can more efficiently clear the lactate from their systems, they may have better endurance and athletic performance.[6]

Another study showed that when sedentary people began to exercise, over the course of six weeks their gut microbes shifted toward more of the bacteria that produce those gut-healthy short-chain fatty acids. When they returned to their sedentary ways, however, their gut bacteria shifted again, this time away from the ones that made SCFAs. What the study seems to show is that the benefits of exercise on your microbiome are real, but last only as long as you keep up the activity. It's a great argument for sticking with your workouts.[7]

Sitting Is the New Smoking

Okay, so now we know that the results last only as long as you continue exercising. But it's also important to understand why a sedentary lifestyle is so bad for you. Researchers recently found that compared to adults who sat only an hour a day, adults who sat for more than ten hours a day had a 34 percent higher risk of death.[8] Another study found that nearly 6 percent of deaths annually in developed countries can be attributed to daily sitting time. For comparison, about 9 percent of deaths every year are from tobacco use and about 5 percent are from being overweight or obese.[9] And to drive the point home, another recent study has shown that the more time you spend sitting and specifically watching TV, the greater your risk of death from all causes. In all the studies, the risks from excess sitting were offset only by a lot of physical activity—more than an hour a day of moderate activity. And if you sat a lot while watching TV, even that level of exercise didn't eliminate your risk.[10] So when you're home watching your shows, don't just sit there—stand up and do your hand weight exercises, stretches, leg lifts—anything to get your butt off the couch. Netflix and Chill is so passé—your newly renovated self needs to Netflix and Move!

Even just standing up from your regular desk more during the day is helpful. Instead of sitting while you take a phone call or a Zoom meeting, stand up. If you're working in an office, have stand-up in-person meetings (bonus: they're shorter), or better yet, have walking meetings. Whether you're working from home or from an office, you can find ways to stand or walk during the day: take the stairs instead of the elevator, park in the far corner of the lot, just stand up and stretch every half hour, or take a walk after eating lunch.

Types of Exercise

Exercise (like ice cream) comes in several flavors, and you need all of them for good health (unlike ice cream, sadly). Regular aerobic exercise, the kind that gets your heart beating faster and makes you breathe harder, improves your cardiovascular fitness and lung function. Good examples are swimming, biking, stair-stepping, and exercises like squats and push-ups. What's the easiest aerobic exercise of all? I'll give you two hints: it's free, and you learned how to do it before you were two years old. You guessed it—a brisk walk. Just put on your walking shoes and go! Another fun option for the home gym is finding online exercise classes and videos. I'll recommend some of my go-tos in chapter 11. And don't forget that any reasonably vigorous activity, like gardening, playing tag with the kids, walking an energetic dog, and even vacuuming the house, can count toward your daily aerobic total.

Weight-bearing workouts are equally important, especially for women, as a key aspect of preventing age-related bone and muscle loss and avoiding osteoporosis—bones that are thin, brittle, and easily broken.

These exercises, which make your body work against gravity, include walking, jogging, running, climbing stairs, and dancing.

Weight training, also called resistance training, is the ideal way to get a weight-bearing workout. It's how you get toned, defined muscles and more muscle mass overall. Increasing your muscle mass boosts your resting metabolism, meaning you burn more calories even when you're sitting still. Gotta love that. You also use your blood sugar more efficiently.

Balance exercises keep you stable on your feet, improve coordination, and help prevent falls (a major concern as we get older). Try to include some balance-enhancing moves in every workout session, or just do them whenever you have a couple of free minutes. I like simple leg stands, where all you do is stand up straight, lift one foot an inch off the floor, and hold it there for ten to fifteen seconds. Repeat with the other leg; do at least five on each side. It's harder than it sounds, but you'll get better the more you do.

Stretching exercises often get forgotten, but they shouldn't be. They help you stay flexible and avoid injury by allowing you to warm up and cool down from more strenuous exercise. Yoga exercises and home Pilates workouts are great for stretching you out and strengthening your core.

YOGA FOR YO GUT

The physical, mental, and spiritual practice of yoga goes back thousands of years to ancient India. Yoga is practiced in a lot of different ways in different spiritual traditions around the world. In modern Western society, yoga is mostly practiced using postures, or asanas, that are primarily stretching and breathing exercises, sometimes in combination with meditation. Yoga can be very helpful for stress reduction; it's also great for improving muscle tone and flexibility.

As an alternative to yoga, or in addition to it, you can try qigong or tai chi. Qigong is part of traditional Chinese medicine. It uses low-impact physical movement, breathing exercises, and relaxation to help you stay physically and mentally balanced. Tai chi grew out of ancient Chinese martial arts. It uses gentle, flowing movements to help with mental focus, relaxation, and physical strength and balance—it's sometimes described as meditation in motion. There's some interesting scientific evidence to support tai chi for stress and pain reduction.[11] To learn qigong or tai chi, you can take in-person classes or check out the many websites with self-teaching videos.

Ideally, you'll do all the different exercise types at least a few times each over the course of a week. Some people like to do a little of each at every workout; others prefer to alternate cardio and strength-training, starting and ending with stretches and doing some balance exercises on most days.

We now know that you can get the effects of a thirty-minute workout in half the time—seriously, only fifteen minutes!—by doing high-intensity interval training, or HIIT. I love HIIT workouts because they're intense but short. You exercise as hard as you can for short bursts of under a minute, then slow the pace down for five minutes or so, then do another burst. Even if you don't need to save those precious extra minutes during your workout, adding some HIIT to your regular routine can help build strength faster. If you're taking a walk, for example, try alternating your regular pace with a couple periods of walking as fast as you can.

Which exercises you do when is a lot less important than just doing them regularly. To help you design your own basic exercise program, I've included some sample exercises and workouts at the end of this book.

HOME OFFICE: STEP AWAY FROM THE ZOOM

During the pandemic, working from home became the norm. For many people, work from home continues, some or all of the time. And while it may seem appealing to wear pajama bottoms all day, there are definitely some downsides to WFH when it comes to your health. Without a commute, many lost the only regular physical activity they had in their day and are now sitting more than ever. People are also working much longer hours. They feel they need to be always online and available, because their bosses know they're at home.

The beauty of being at home is that you have the power to create a healthier work environment. One simple renovation step: invest in a standing desk. Multiple studies have proven the many health benefits of using these desks, and they actually have been shown to make you more productive at work.[12]

Muscles and Your Microbiome: Stay Strong

As we age, we naturally lose lean muscle mass. Some sarcopenia—age-related loss of muscle mass and function—is inevitable. Even if you're active, you'll still have some muscle loss starting in your forties. But the amount you lose depends mostly on how hard you fight back with regular strength-building exercise. Working out with weights a few times a week is a key strategy for staying strong and protecting you against some of the problems of aging.

When your thigh and butt muscles are strong, for example, you're less likely to develop painful knee arthritis. Strong muscles can also help prevent falls that can cause injuries such as concussion or broken bones in older adults. And from an aesthetic perspective, toned and defined muscles are your ticket to preventing or reversing sagging upper arms and droopy other parts and stubborn belly fat.

Another factor in building strong muscles is protein. As you get older, you actually need more than you used to.[13] Take your weight and divide it by two, then aim for that amount in protein grams each day. To put your protein needs in perspective, a chicken leg has about 12 grams of protein; an egg has about 6 grams; a 5-ounce can of light tuna has 20 grams.

In addition to exercise, you may also be able to fight back against muscle loss with your bacteria. The few studies on this question focused mainly on older adults who were frail and had limited activity due to reduced muscle strength. In one study, a group of frail people added inulin, a type of fiber, to their diet. They didn't exercise more, but their hand-grip strength improved. The inulin likely shifted their gut bacteria in the direction of producing more short-chain fatty acids.[14]

Yes, it would be great if you could just take a probiotic pill with the right bacteria and never work out again, but hold your horses, we're not there yet. What the research is telling us so far is that a healthy gut microbiome can help maintain muscle mass, muscle function, and good physical performance as we age, but only if you also stay fit.

Bones and Your Microbiome

By the time you're thirty, your bones have reached their maximum density and strength. That means by the time you're forty, your bones have slowly started to lose some density. If you're a woman, the process accelerates when you reach menopause and lose the protective effect of estrogen, which means you're at a greater risk for osteoporosis. A fall you shook off when you were thirty could break a bone when you're sixty. Men also lose bone density with age, although less so because their bones are bigger and heavier.

Although some bone loss with age is inevitable, weight-bearing exercises can slow it down and prevent osteoporosis. You can probably guess the next thing I'm going to say: your gut bacteria play a role in bone strength. Correct! How much of a role is becoming increasingly apparent. One recent study in Sweden suggests probiotics can help improve bone density in postmenopausal women. The study looked at ninety older women (average age seventy-six) who had low bone density. Half took a probiotic supplement for a year; the others took a placebo. At the end of the year, the women in the probiotic group had lost only half as much bone as those in the placebo group. Nobody in the probiotics group had any side effects from taking them.[15] This last point is more important than it sounds, because the commonly prescribed drugs to slow bone loss are hard to tolerate and can have some serious side effects.

As a preventive measure and treatment for osteoporosis, probiotics show a lot of promise. But yes, you still have to exercise!

Joints and Your Microbiome

Osteoarthritis (OA), more commonly just called arthritis, is the most common musculoskeletal disease in the world—almost everyone over the age of sixty has at least some mild arthritis. If you have OA, the cartilage that normally cushions your joints has worn down. In other words, your joints hurt.

You're more likely to get severe osteoarthritis if you're older, female, overweight or obese, have a poor diet with lots of processed foods and sugar, and are physically inactive. We've now added inflammaging to those risk factors. In fact, it's probably inflammaging that kicks off the process that gets arthritis started. And what causes the inflammaging? The chronic, systemic, low-grade inflammation caused by dysbiosis.[16]

In my practice, I see a lot of patients with rheumatoid arthritis (RA), an autoimmune condition that affects the joints. Why are they in my gastroenterology office? Because RA can affect the gut as well as the joints. In fact, more and more research suggests that an imbalance among the types of bacteria in the gut microbiome is actually an underlying cause of RA. One clue is that people with RA are usually treated with disease-modifying drugs that suppress inflammation by targeting the cascade of cytokines produced by immune cells. At the same time, most of these drugs turn out to have a positive impact on the gut bacteria. It's possible that people who take them show improvement because the drugs help restore a better balance of bacteria in the gut.[17]

On the flip side, I often refer my patients with inflammatory bowel disease, such as Crohn's disease, to a rheumatologist. Why? Because IBD often causes joint problems. Here too, the link between inflammatory gut problems and joint problems is the microbiome. People with inflammatory bowel disease often have a leaky gut. Bacteria from the gut, along with bacterial waste products and metabolites, escape into their circulation, causing inflammation in the rest of the body. All too often, the inflammation targets the joints and causes pain and swelling. In fact, about 20 percent of people with Crohn's disease also have arthritis from it. The arthritis gets worse when they have a disease flare-up and tends to improve when the disease improves. For most, aching joints are a manageable part of the disease. Some will develop a severe form of arthritis called ankylosing spondylitis, which affects the spine and the sacroiliac joints (the joints that connect your pelvis to your spine).

You can't do anything about getting older, but it's amazing how much is in our control. A healthy Mediterranean-style diet, along with prebiotic and probiotic supplements, will improve the gut bacteria composition and help reduce the inflammation that both causes and worsens osteoarthritis. And exercise is your secret weapon! Even getting in a couple of brisk walks every day will strengthen your muscles and relieve stress on your joints.

So now that you've worked up a sweat learning how to renovate your home gym, it's time to move on to the next chapter, where you'll get a well-deserved break, a space to cool down and chill out—your Zen Corner.

GUT TOOLBOX

- *Hop in line.* As someone who hates waiting in lines, I now tolerate them much more because I use that time to practice my balance moves. You may raise a few eyebrows if you try this, but it's a perfect opportunity to work on your balance—and in my case, patience. Try it out!

- *Find a mat mate.* Working out or going to an exercise class with a friend is always more fun and motivating. Making this a standing date keeps you consistent and accountable.

- *Find fitness funding.* If you work for a company, you may find that they subsidize gym memberships or other wellness products. Many companies offer corporate fitness challenges or benefits, so explore your options!

- *Do push-ups to podcasts.* If you find exercising boring, then add some enjoyment with a favorite song playlist, audio book, or podcast. Time really does fly when you're having fun.

CHAPTER 7

THE ZEN CORNER

Now you know how physical activity is a key part of your Gut Renovation, as is staying mentally active. But what if your mind is *too* active? Racing from one stressful thought to the next?

Well, I can tell you that's pretty common.

But that doesn't *need* to be the way of things. We all need strategies to quiet the noise in our heads, which is why the next space we'll discuss is an essential part of your Gut Renovation: your Zen Corner.

In a literal sense, making a Zen Corner in your home is probably the easiest and cheapest renovation you can make. You don't need a lot of space, because this corner can go anywhere—the bedroom and living room are popular choices. It doesn't even need to be a corner! All you need is a quiet place big enough for a yoga mat or comfy cushion (and preferably a huge "Do Not Disturb" sign).

In your Gut Renovation, the Zen Corner is the space where you can focus on your mental health—and you'll see how much that impacts your microbiome, and vice versa! Wherever and however you set up your Zen Corner, the most important thing to remember is that finding ways to pause, and deal with stress or boost your mood, will greatly benefit your health in a variety of ways.

What Is Stress?

Stress is an all-encompassing word we often use to describe our feelings about the challenges we face. Sometimes the stress is short-term, based on everyday issues, like worrying about a speech you have to give. Sometimes it's longer-term, like when you're going through a break-up. And sometimes stress is chronic, like constantly dealing with difficult emotions, or living or working in a toxic environment, or worrying about finances.

Chronic stress means your autonomic nervous system is always making your adrenal glands release stress hormones, especially cortisol. You're always in the on state of fight-or-flight, which means your body doesn't get to relax. As that suggests, chronic stress can really mess up your digestive system. Not only do the stress hormones cause nausea, vomiting, diarrhea and constipation, and other digestive symptoms, they can make you hypersensitive to gut sensations, like abdominal pain. And when your cortisol levels crash, that can cause food cravings, especially for sugar and fatty foods. Even short-term stress can have a negative impact on your gut microbiome; chronic stress can cause dysbiosis, increased intestinal permeability, and a leaky gut, with all the digestive problems they bring—and the impact they have on your mood and emotions through the gut–brain axis.

Microbes and Mood

My patients with irritable bowel syndrome and other chronic inflammatory problems are likely to feel worse digestive symptoms when they go through a period when they feel anxious, depressed, or stressed. They have flare-ups of symptoms, more inflammation, and can feel more overwhelmed by the challenges of managing their condition.

What my IBS patients experience also happens, to some degree, to most of us at some point. Life happens, sometimes causing anger, worry, irritation, or stress. Whatever it is, it's probably affecting your microbiome. At times it may be the other way around—your microbiome is actually affecting your emotions and your mood.

We're just starting to unravel the links between your gut microbiome and your brain processes. They happen at a lot of levels—through your vagus nerve, your enteric nervous system, your hormones, your immune system, your neurotransmitters, through the effects of a leaky gut, and through the signals your microbial metabolites send. (For more details on the gut–brain axis, circle back to chapter 2, "The Architect.")

Changes in your microbiome that affect the proportions of different bacteria families can wind up having a profound impact on your brain. Even a brief period of stress can change your gut microbiome profile. Take your gut's production of serotonin, for example. You already learned how specialized cells in your gut lining produce large amounts of this neurotransmitter when they're stimulated by the presence of some types of gut bacteria. When you feel stressed, changes in your food intake (eating way too much ice cream, for instance) can alter the composition of your gut bacteria. Add in the impact of stress hormones such as cortisol on your gut bacteria and your microbiome changes even more, affecting how much or how little serotonin you can produce. Serotonin is key to how quickly food travels through your gut. Too much serotonin, and the food moves through too speedily, giving you diarrhea. Too little, and the food moves too slowly, giving you constipation.

Stress can also affect every aspect of your digestion. Performance anxiety, aka stage fright, is well-known for activating your sympathetic nervous system and flooding your body with the hormone adrenaline. Your heart races, your mouth gets dry, you break out into a sweat—and might get nauseous and even throw up.

Looking at it from the other direction, your gut microbiome can influence your mood. For example, dysbiosis can cause a leaky gut, which in turn causes systemic inflammation. What's one well-known symptom of inflammation? Feeling tired and depressed.

The exact mechanisms for how your gut bacteria communicate with your brain and how they affect your mood and emotions are still being actively researched, but I find the information that is emerging to be fascinating. The field of nutritional psychiatry, for instance, has some good evidence for some dietary interventions in people with severe depression. They probably help, at least in part, by causing favorable changes in the gut bacteria of these people. One recent study showed that taking a daily prebiotic supplement for four weeks reduced anxiety in young people.[1] And another trial demonstrated that in women with self-reported anxiety, one month of a prebiotic supplement not only lowered anxiety levels but also improved their microbiome and brain health, based on microbiome sequencing and brain imaging.[2]

That's still a long way from prescribing a particular probiotic strain to treat severe depression. I do see a day, maybe soon, when mental health professionals start paying a lot more attention to the potential of a better diet and probiotics as tools to help their clients.

What we can say is that a gut-healthy diet with plenty of fiber, prebiotics, and probiotics helps keep your gut bacteria happy and in balance with each other. That in turn can play a role in keeping your mood and emotions on an even keel.

Leaky Gut, Leaky Marriage

A fascinating example of how your emotions affect your gut comes from recent research on couples who fight. Okay, all couples fight, but we're talking about next-level fights. It turns out that if you and your partner have frequent nasty arguments, you're both more likely to develop leaky gut syndrome. In the study, forty-three couples had blood samples taken before being asked to try to resolve a conflict in their relationship that the researchers deemed likely to provoke strong disagreement. Additional blood samples were taken after the argument ended. The researchers found the couples who showed the most hostility during the argument also showed higher levels of a biomarker for leaky gut that indicates bacteria in the bloodstream. They also, not surprisingly, tested higher for markers of inflammation. The chronic stress of being in a bad intimate-partner relationship is clearly bad for your digestive health, as the leaky gut marker showed, but also bad for your overall health, as the inflammation markers showed.[3] Not that I'm advocating mass break-ups, but trying to find less stressful ways to resolve conflict is important. In my next book, *Gut Renovate Your Marriage* . . . just kidding. But seriously, a great resource for healthier arguing and a healthier relationship that I think should be required reading for any couple is the website of the Gottman Institute at https://www.gottman.com.

Food and Mood

When you're feeling down, you might decide to settle in with a pint of Rocky Road for an evening of movies. Why do you choose ice cream instead of, say, carrots? In part, it's because your comfort food of choice may trigger an emotional release—ice cream, for example, takes you back to childhood, when life was simpler (and okay, yes, ice cream is delicious). You're also unconsciously using food to influence your mood. The sugar, fat, and tryptophan from the milk all have a temporary calming effect on your brain's neurotransmitters. But long-term sugar intake can cause just the opposite: an increased risk of anxiety or depression. And if you chronically eat large amounts of sugar and then stop, sugar withdrawal can lead to low mood and irritability.

Your ability to think clearly and stay on an even emotional keel are profoundly affected by what you eat. The macro nutrients (protein, carbohydrates, fats) in your food, along with the micronutrients (vitamins, minerals, phytonutrients) affect your production of neurotransmitters, hormones, and enzymes, both in your gut and in your brain. Not enough or too much of any of the building blocks will throw off their production—and could throw off your mood, your cognition, and even your sleep. As one example, you need B vitamins for good brain function. Vitamin B6 (pyridoxine) is needed to make dopamine (another happiness hormone), and you need vitamins B6, B9, and B12 to make serotonin. Not enough Bs to make these neurotransmitters and you feel depressed and uninterested in life. For some people, taking B vitamin supplements really helps lift their mood.

Tyrosine, found in fish, nuts, eggs, beans, and whole grains, is another key component of dopamine. Since dopamine plays an important role in regulating your mood, alertness, and ability to learn, you can see how adding more tyrosine to your diet could affect your state of mind.

In the short term, stress usually damps down your appetite by releasing the hormone epinephrine, which puts your appetite on hold. Long-term constant stress, however, is a trigger for the hormone cortisol, which revs you up in a lot of ways, including your appetite (possibly by increasing production of the hunger hormone ghrelin). You feel a strong motivation to eat—stress eating. Mostly you crave sugary and fatty foods, because your body knows that they will counteract the stress and make you feel better in the near term. Intellectually, you know that these foods are bad for you, but your body overrules your brain.

I understand that conflict all too well, because I see the impact on my patients. People who were doing well at controlling their gut symptoms through a good diet encounter a rough patch in their life, end up stress eating, and then end up back in my office.

That's when we have a serious discussion about stress management.

Learning to Manage Everyday Stress

One of the biggest parts of the Zen Corner is handling the issues that come up and can stress us out on any given day.

Now, what works for you may not work for someone else, but you're the one that counts.

Your Gut Renovation includes incorporating DIY steps to handle everyday stress.

One of the simplest ways to relieve stress in the moment is to give yourself ten quiet minutes to relax. What better way to do that than with a hot cup of relaxing tea? In traditional herbal medicine, chamomile, peppermint, lemon balm, lavender, and rosehip teas are all recommended for relaxation. Is there science behind the relaxation effects? Sort of, but in this case it doesn't really matter. Just find a flavor or blend you like and relax with the ritual of preparing it and then sipping it in a quiet place.

If a cup of tea isn't convenient or really won't help with the sort of stress you're dealing with at the moment, try deep breathing. You can do this anywhere, anytime—it only takes a few minutes to have a strong impact. Simply breathe in deeply through your nose for four seconds, hold for a count of five, then breathe out slowly through your mouth for six seconds. Repeat at least five times.

This works because when you're stressed and anxious, you tend to take shallow breaths. Deep breathing gets more oxygen into your body, which helps calm you down and get you refocused. Anxiety and stress activate your sympathetic nervous system—the fight-or-flight response. When you breathe deeply, you relax and activate your parasympathetic nervous system, the rest-and-relax response. Your body can't have both responses together.

Even just getting touched by someone supportive can help relieve stress. A big hug, a shoulder rub, holding hands, even just a pat on the back brings down cortisol levels and increases levels of oxytocin, the "love hormone." Supportive touch also sends messages up the vagus nerve telling the brain that things are okay; the message back tells your body to relax. In one study, just holding hands with an intimate partner reduced the sensation of pain when a mild shock was applied to one person in the pair.[4]

The Media Room

These days, our use of media spills way beyond one room in our home. Let's face it, there's no longer a special TV room—we have devices everywhere! In fact, for many of us, our phones appear surgically attached to our hands for most of our waking hours. We end up checking them constantly for messages and email, looking at our social media all the time and worrying about our likes and comments, staying up late scrolling. Then there's video games, all those TV channels, and computer games. All of that digital distraction can be stressful, and even addictive. A 2018 study of college students found that the less time they spent on social media, the less depressed and lonely they felt.[5]

Which is why a crucial renovation tool in your Zen Corner is the daily digital detox. Turn off your smartphone, power down your tablet, put your computer to sleep, switch off the TV for at least an hour a day, preferably before bedtime. Spend that time making "in real life" connections with your family and friends, reading a printed book, or doing something you enjoy, like a craft. Even better, spend that time on one of the healthiest practices around: meditation.

Meditate on This

Meditation is the practice of focusing your attention and awareness as a way to reach a mentally clear and emotionally calm state of mind. Prayer, for instance, can be a form of meditation. In our secular society, however, meditation usually means using a technique such as mindfulness or Zen meditation as a way to reduce stress and increase calmness and feelings of well-being.

Mindfulness is the practice of setting aside time to pay attention in the moment to your thoughts and feelings. When you increase your awareness of what you're thinking and feeling, you're better able to see them in perspective, as part of your larger environment. Getting some perspective in this nonjudgmental way can really help you realize where the stressful areas of your life are. By noticing and acknowledging signs of stress in your life, you take away their power to make you worried, anxious, or depressed.

Mindfulness techniques are usually based on the idea of focusing your attention on something, such as your breath, a mantra, a visualization, or even an object, as a way of calming your mind and centering your thoughts. Plenty of books, apps, websites, videos, and workshops that teach mindfulness are available, but you don't need any special training to learn how to do it. All you need is a quiet, comfortable place to sit for ten to twenty minutes without being interrupted. Sit quietly and focus your attention on your breathing. When you first begin your mindfulness practice, you'll find it hard to stay focused. You'll have lots of stray thoughts; worries will keep intruding. That's perfectly fine. Just calmly notice those thoughts as they enter and then exit your consciousness, and then return to your focus. After a while, you'll find you're able to let go of those distractions, enter the mindfulness zone, and stay there easily.

Mindfulness has become very popular because it's a great way to regulate your feelings and help with unhealthy patterns. We need this now more than ever. The prolonged stress of the COVID-19 pandemic has led to a lot of lousy coping mechanisms, like increased drinking and overeating. In fact, polling by the American Psychological Association in 2021 showed that 61 percent of adults experienced undesired weight change—usually gaining—during the pandemic. The average reported weight gain was twenty-nine pounds! Nearly one in four adults reported drinking more alcohol to cope with stress.[6] Mindfulness and meditation can help reduce the stress causing these unhealthy behaviors.

The key to an effective mindfulness practice is doing it every day, or at least as regularly as you can. The stressful things in your life will still be there; what will change is your ability to see them for what they are and cope with them better.

Just as mindfulness and meditation can help reduce stress and make you feel calmer and happier, it can do the same for your microbiome. Many studies consistently show that stress hormones disrupt your microbiome's production of short-chain fatty acids and other anti-inflammatory chemicals; it also affects the production and regulation of neurotransmitters and makes the gut barrier more permeable. When you meditate regularly, you regulate your stress response and reduce the amount of stress hormones you produce. That lowers chronic inflammation and restores gut barrier function.[7]

Gastroenterologists for decades have recommended stress reduction to their patients with irritable bowel syndrome. We knew it often worked to help reduce symptoms, even though we didn't have a lot of research data to back it up. In 2017, a solid study showed through measurable outcomes, not just patient self-reports, that mindfulness can indeed improve symptoms. The study looked at seventy-five women with IBS. Half were randomly assigned to eight weekly mindfulness training sessions; the other half attended a support group. Three months later, the patients who completed the mindfulness training showed real reductions in their symptom severity and had better quality of life, less psychological distress, and less visceral anxiety (disproportionate anxiety related to GI sensations). The women in the support group showed far less improvement after three months.[8]

Other Ways to Meditate

Meditation is an expansive idea—I consider it basically anything that helps focus your attention, calm your thoughts, and relax. The important thing is to do it regularly so that you can easily enter into the meditative state. Even knitting can be a form of meditation. For some people, just having some quiet alone time to read a book is enough.

I suggest trying out some meditation exercises to find one or more that really work for you. Guided imagery, where you sit quietly and visualize a place or situation you find relaxing, works well for some people. Really imagine your happy place, down to the way it smells and feels.

Another great exercise is a gratitude meditation. You visualize all the things in your life you're grateful for. This is a great way to get perspective on the stressful aspects of your life. When you see the stressors relative to all you have to be grateful for, they matter a lot less. Not only does this approach help with depressive symptoms, but it's also been shown to lower inflammatory markers and stress hormones.[9]

Candle meditation is another easy change of pace (and a great way to use one of those scented candles someone gifted you). All you have to do is turn off or dim the lights, sit comfortably, and put your lit candle in front of you. Focus your gaze on the flickering candle flame and be aware of your breathing. If your thoughts wander, just refocus. Continue for about five minutes.

The Backyard

In addition to all the positive health effects of walking that we discussed in the "Home Gym" chapter, walking is also a great form of meditation. A quiet walk in a pleasant environment is relaxing in itself. When you focus on the rhythm of your gait as you walk, just as you focus on your breathing during a sitting meditation, you can redirect your thoughts away from the stressful things around you and be in the moment. Ideally, you take the walk outside. Spending time outdoors in nature is so good for you that the Japanese have a word for it: *shinrin-yoku*, meaning "forest bathing." Spending time among trees or green spaces also has proven health benefits, including boosting your immune system, lowering your blood pressure, and improving your ability to concentrate.[10]

Positive Self-Talk

Your self-talk is your internal dialogue, the conversation you're constantly having with yourself. What you say when you talk to yourself can be very revealing. An important step in your Gut Renovation is to toss out the self-criticism. Negative self-talk, where you criticize and question yourself, contributes to a pessimistic outlook on life, anxiety, and depression. Negative self-talk often turns into an endless loop of toxic thoughts and worry. Positive self-talk, where you encourage and support yourself, is a big part of having an optimistic outlook on life and managing stress successfully.

When you hear your inner voice stuck on a negative train of thought, one technique to redirect your brain is to visualize a bright red stop sign. This vivid picture can signal your brain to STOP obsessing over something distressing and give you a chance to think of something more positive. Self-affirmation is another important tool. As cheesy as it sounds, every morning when you look in the mirror, give yourself one compliment. Say it out loud. It can be small, like "You have nice eyebrows," or bigger, like "You're a good listener." Another good one, especially if you're going through a hard time, is "You're doing your best."

Finally, a word about therapy. You can't spend twenty years as a gastroenterologist without appreciating the benefits of it! I've seen therapy help countless patients, with both their gut health and their mental health. For many of my patients, cognitive behavioral therapy (CBT) has been a real help for learning to deal with their symptoms and reducing the stress that makes those symptoms worse. CBT is psychological therapy that helps you recognize negative behavior and thought patterns and learn to reframe them in a more positive way. CBT is based on the idea that our own thoughts cause our feelings and behaviors—and that means you can change your thinking and feel better, even if the situation doesn't change. That's why CBT is so great for my IBS and IBD patients. Therapy can help them be more accepting of their illness and learn to focus their intelligence and energy toward managing it successfully. When patients start to feel they're in control of their illness instead of the other way round, their stress levels go way down, and their symptoms usually improve. What I like about CBT for my patients is that it works quickly, and the lessons stick with them—they get long-term results. I also like the scientific evidence that backs up CBT for digestive problems. A number of high-quality studies have looked at CBT, particularly for IBS, and found that it's effective.[11]

There are many different kinds and many different situations where therapy can be useful, and the truth is, one of the smartest and strongest decisions you can make is to do therapy with a mental health professional. In fact, a girlfriend told me the other day that she doesn't trust men who have *not* gone to therapy. So if you think you may possibly benefit, go for it. I've done it myself and I've never regretted it. Having a trained, nonjudgmental helper in your Zen Corner may be just the renovation tool you need.

GUT TOOLBOX 

- *Silence, please.* It's impossible to relax when your phone is constantly beeping or buzzing. Turning off all notifications except the truly essential ones keeps you from being disturbed and potentially stressed out every five minutes. Instead, you can set regular times to check your various apps, rather than constantly being drawn to them.

- *Find your raisin d'etre.* For a change of pace in your usual meditation practice, try the raisin meditation. Put a single raisin in front of you and imagine you've never before seen or tasted a raisin. Focus your attention on the raisin, exploring every aspect of it. How does it look, feel, smell? Then eat the raisin, chewing very slowly. How does it taste? You're using the raisin (or any other convenient small piece of food) to focus your attention away from the stressful things around you. This is a form of mindful eating, where you slow down your eating pace and pay close attention to each bite.

- *Go to the dark side.* One study showed that eating just 1.4 ounces of dark chocolate every day for two weeks reduced levels of cortisol and another family of stress hormones called catecholamines in highly stressed people.[12] Dark chocolate has also been shown to have a prebiotic effect on the microbiome, promoting the growth of beneficial bacteria.[13] Good for your gut, good for your stress, and definitely good for your taste buds!

- *Stick it out.* Placing a motivational sticker on your wall or mirror that has a self-affirming message can do wonders for your daily mood. It may sound silly, but trust me, it works.

- *Be your own bestie.* Distanced self-talk, where you talk to yourself as if you were talking to another person, can be very helpful for breaking a negative loop and getting some perspective on your thinking. We're often much harsher on ourselves than we would be to a friend if they were in our situation.

CHAPTER 8

THE BEDROOM: SLEEPING YOUR WAY TO BETTER HEALTH

If all that mellow Zen talk put you in the mood to lie down, then you're in the perfect spot—your bedroom. This is where you spend about a third of your day, so it should be one of the most important elements of your Gut Renovation.

Thanks to the sleep revolution, we know now just how critical rest is for the body—and yes, I'm talking specifically about gut health. Get better sleep and you'll have better digestion and gut health. And vice versa—your hard-working general contractor doesn't go off duty at night. The microbiome-sleep connection is real and can be used to your advantage.

But even though we know how important sleep is, a huge proportion of us only dream (pardon the pun) of getting a good night's sleep. That number increased even more during the pandemic, where the stress caused by isolation, working from home, financial insecurity, and the disruption of routines led many to develop "coronosomnia."[1]

You *Don't* Snooze, You Lose

When life gets busy and there just aren't enough hours in the day, the first thing you're likely to jettison is sleep. You rationalize that by saying you'll go to bed early the next night, or sleep late on the weekend, or say, "I'll sleep when I'm dead." I hate to break it to you, but if you don't get enough sleep, that might happen sooner than you think.

Adults generally need between seven and nine hours of sleep each night. The amount varies among individuals—some lucky types are short sleepers who are fine with under seven hours a night; some need more than nine hours to feel fully functional. You likely already know how long you need to sleep to wake up feeling refreshed and ready for the day—and that you're probably not reaching your ideal sleep number most of the week. At least one in three adults don't get enough uninterrupted sleep on a regular basis, and about fifty to seventy million Americans have sleep disorders such as sleep apnea or chronic insomnia.[2]

The broad term for all this lack of sleep is sleep deficiency. It occurs when you don't get enough sleep (sleep deprivation), but it also happens when you get out of sync with your circadian rhythm and end up sleeping at the wrong time of day. Not having good-quality sleep or not getting enough of all four sleep stages is another common reason for sleep deficiency. And then there are sleep disorders that regularly prevent you from getting a restful sleep.

Big picture, lack of sleep is linked to an increased risk of heart disease, kidney disease, high blood pressure, diabetes, stroke, obesity, cognitive impairment, and some kinds of cancer, to name a few.

Another example of how poor sleep affects your health is your metabolism. If your sleep pattern is consistently irregular—you don't go to sleep at approximately the same time each evening and wake at approximately the same time the next morning—you're a lot more likely to have a metabolic disorder such as obesity, high blood sugar, type 2 diabetes, and high cholesterol. A recent study showed that for every hour of variability in the time you spend in bed and asleep, your risk of developing a metabolic disorder increases up to 27 percent.[3]

Another important role of sleep we're just beginning to understand is how it relates to cognitive decline as we age. A number of studies have linked poor sleep or sleep deprivation among older adults to a greater risk of dementia and death, compared to older adults who get more and better sleep. The risk is even worse for those who regularly sleep less than five hours a night.[4] While you're deeply asleep and not dreaming (Stage 2 and Stage 3 sleep), brain immune cells called microglia get busy taking out your brain's neural trash. They remove the accumulated toxins of the day, which helps keep the protein plaques and tangles of Alzheimer's disease from forming. At the same time, your sleeping brain lets waves of the fluid that bathes your spinal cord and brain flush out the toxins. When you don't spend enough time in the deep sleep stages, your brain can't clear out the garbage well enough and your risk of neurodegenerative disease goes up.[5]

Why You Sleep

Sleep is just as essential to your health as food and water. In fact, your body does a mini-renovation every night when you sleep. You may be snoozing, but your body is very busy as it repairs or removes damaged cells, releases hormones for growth and repair (and also hormones related to hunger and appetite) and builds up energy levels for when you're awake. In your brain, sleep helps you make long-term memories and process and store new information. Sleep is also key for efficient executive function when you're awake: you need to be well rested to focus, solve problems, concentrate, make decisions, and regulate your emotions.

Your sleep–wake cycle is governed by two internal biological clocks: your circadian rhythm and your sleep–wake homeostasis. Your circadian rhythms are physical and mental processes that follow a twenty-four-hour cycle responding to light and dark. Sleep homeostasis is the drive to sleep, based on how long you've been awake. Over the course of roughly twenty-four hours, these two internal biological clocks interact and tell your body when to sleep and when to be awake—and also control other functions, including body temperature, hormone release, and your metabolism. Sleep–wake homeostasis works in sync with your circadian clock to track how long you've been awake and to tell you it's time for bed. The homeostatic sleep drive builds up and gets stronger every hour you're awake, until you finally give in and go to bed. Both internal clocks respond to environmental cues, especially light, to make you wake up naturally in the morning. They also both contribute to the natural ebb and flow of your alertness over the course of the day.

Most people have a sleep–wake cycle that syncs nicely with the twenty-four-hour day–night cycle. However, lots of factors can mess up your sleep–wake cycle, including stress, shift work, jet lag, illness, pain, some medications, your sleep environment (a snoring partner, for example), and what you eat and drink in the hours before you go to sleep.

Melatonin and Your Sleep

Your body's circadian clock gets reset every morning by the body's central clock, a tiny clump of brain cells behind your eyes called the suprachiasmatic nucleus (SCN). As the sun goes down and you've been awake for hours, the SCN senses darkness and signals the pineal gland in your brain to start secreting the hormone melatonin. This hormone is produced only during darkness and stops when you're exposed to bright light.

Melatonin from the pineal gland regulates the timing of your sleep–wake cycle. As sleep pressure builds in the evening, melatonin gets your body's systems and processes synced up for sleep and makes you feel drowsy. The amount of melatonin in your bloodstream gradually builds, reaching a peak between two and four in the morning, and then gradually decreases. As the sun comes up, your body stops producing melatonin and starts releasing the hormone cortisol to prepare you to wake up.

Do melatonin supplements help you get to sleep? They might, if you take a high-quality product at just the right time. Look for products from reputable manufacturers. The label should have the USP or NSF International seal, along with a current good manufacturing processes (cGMPs) statement. Take the supplement about two hours before your bedtime—the same time that your body is naturally starting to produce melatonin. The usual dose is 1 to 3 milligrams.

Slumber Party in Your Gut

Just as you have a circadian clock, so do the bacteria that live in your gut. Your gut microbiome's circadian cycle is linked to your sleep–wake cycle and also to the rhythm of your eating pattern. When your sleep–wake cycle is disrupted, your eating pattern usually is as well. Here's how it works: You need adequate sleep to regulate your hunger hormones, especially the hormones leptin and ghrelin. When you don't get enough sleep, you trigger increased levels of ghrelin, which makes you hungry, and decreased levels of leptin, which tells you you're full. You feel hungry all the time—and because you're awake so much, you have lots of extra time to overeat.[6] Not only that, but lack of sleep also stimulates cravings for high-calorie, sugary, and fatty foods and makes us see those foods as a reward.[7]

Raise your hand if you can relate to this. I know I can. Back in my residency days, when staying up for twenty-four to thirty-six hours was the norm, my go-to breakfast after a long shift was a huge egg, sausage, and cheese sandwich on a croissant. My exhausted body craved lots of food, and unfortunately not the healthy kind.

When you're sleep deficient, you tend to eat a lot of high-calorie, low-fiber junk food, and you eat them outside of your usual mealtimes. Your microbiome doesn't like that—and it may let you know by giving you diarrhea and/or constipation, gas, or bloating. The impact of disrupted sleep on your microbiome is so familiar to health care workers, emergency personnel, and others that it has a name: night-shift belly. You can get over the digestive upsets caused by a week or two of night shifts, but over time, the disruption caused by too little sleep or too much disturbed sleep can cause dysbiosis, systemic inflammation, and possibly leaky gut syndrome.[8]

Frequently being short of sleep can also lead to weight gain from cravings and from stress overeating, which can lead to additional conditions that can disrupt your sleep even more, like sleep apnea and GERD.

In my practice, I see many patients whose digestive problems significantly improve when we work on their sleep. Any digestive issue that gets worse with stress—like IBS, for example—gets better with improved sleep because better sleep means lower production of stress hormones. Better sleep also really helps with gut issues related to your microbiome and inflammation. Indirectly, it helps because when you sleep better, your diet is healthier (more gut-friendly) as well. And when the circadian rhythms of your sleep are in sync with the circadian rhythm of your gut, your digestive symptoms tend to calm down.

Lack of sleep affects your microbiome more directly as well. About 60 percent of your gut microbes change in response to the light–dark circadian rhythm, some species increasing in number and some decreasing. That means that your exposure to the different gut bacteria and their metabolites fluctuates throughout the day. This is an interesting area for researchers, because it could help explain why some people get insomnia and why in some conditions, the symptoms of an illness get better and worse over the course of a day. For example, people with severe depression often feel worse in the morning and better at night.[9] Some early studies have also looked at the effects of probiotics on sleep, and the results are promising.[10]

Your Digestion and Sleep

Just as poor sleep can lead to digestive problems, digestive problems can lead to poor sleep. Many of the patients I see have acid reflux or gastroesophageal reflux disease (GERD). Occasional heartburn that keeps you up at night happens to almost everybody. It's the price you pay for going out with the gang for taco-and-tequila Tuesday. You can generally deal with that sort of heartburn with over-the-counter remedies or even just a good burp. But GERD is another story—it can keep you up night after night and make you seriously sleep-deprived. GERD gets worse when you go to bed, because when you lie down, gravity doesn't help keep acid down in your stomach where it belongs. The acid is more likely to travel up into the esophagus where it can cause burning and discomfort.

In addition to diet changes for these patients, I also recommend DIY steps that can lead to better sleep. Step one is to avoid eating for at least three hours before bedtime to keep food in your stomach from pressing up against your esophageal sphincter (which lies between the stomach and esophagus) when you lie down to sleep. That really helps cut back on the reflux of stomach acid into the esophagus when the sphincter gives way. And skip the foods (and alcohol) you know give you GERD! Lay off those fried mozzarella sticks! (Check back to chapter 3, "The Kitchen," for more.)

Step two is to raise the head of your bed by at least six inches. You have to elevate the bed itself; just using a pillow or foam wedge to elevate your head won't help (and you might wake up with a stiff neck in the morning).

Step three is to sleep on your left side—this helps reduce reflux episodes by putting pressure on the upper part of your J-shaped stomach and keeping the contents from pressing against your lower esophageal sphincter.[11]

The other way GERD can affect your sleep is if it's associated with obstructive sleep apnea, a disorder that makes your breathing repeatedly get very shallow or even briefly stop while you sleep. The two conditions are found together about 60 percent of the time. As a gastroenterologist, I treat the GERD part of the combination. Many patients find improvement in their sleep apnea when the GERD is well controlled, and vice versa.

Jet Lag

Jet lag happens when your circadian rhythm doesn't sync up with the local day–night pattern. It's a common problem for people who travel across time zones. We think of it as a sleep–wake problem and a cause of insomnia, but as you probably have experienced, jet lag can also really mess up your digestion. Jet lag has been linked to dysbiosis, probably due to eating pattern changes, but also from interruptions of the gut microbiome's circadian rhythms.[12] How does this manifest? Feeling hungry at strange times or perhaps an urgent need for a bathroom at an unexpected time, both of which may mess up your sleep even more. To avoid the worst digestive symptoms, be sure to stay well hydrated during your flight. Before you board, skip the airport fast-food chains and find someplace with a good salad bar. Drink plenty of plain water and skip the caffeine and alcohol. Once you've arrived, you can limit the jet lag damage to your gut by continuing to avoid heavy meals, excess caffeine, and alcohol. I've found that exercising outdoors during the day helps reset my internal clock to local time a bit faster. Taking melatonin supplements can also help you reset, but bear in mind that melatonin isn't a sleeping pill. Its role is to get your body systems set up for sleep to begin. A lot of melatonin products are very low-quality—they may work more by the placebo effect than by giving you any meaningful amount of the hormone.

Insomnia

Everyone has the occasional night where they just can't sleep. Maybe it starts by worrying about something, which leads to tossing and turning. Maybe at some point you manage to fall asleep but wake up shortly thereafter. In the morning, you feel tired and unrefreshed and may struggle to get through the day. But the next night, you go to sleep at your usual time or even a bit earlier and sleep well. This is what is known as acute, or short-term, insomnia. It usually lasts for just a night or two, but it can go on for longer. When insomnia happens at least three days a week for at least a month, it's classified as chronic.

Chronic insomnia can go hand in hand with gastrointestinal problems. About a third of people with gastrointestinal problems say they have chronic insomnia.[13] Discomfort from a digestive condition, combined with anxiety about it, can really impact sleep and take a toll on the microbiome. I work with my patients to get the pain and discomfort (and the stress they cause) under control so they can sleep more soundly. I rarely prescribe sleeping aid medications—they work for only a short time, and they have some significant side effects. For long-term insomnia, working with a CBT therapist with special training in insomnia can be very helpful (check back to chapter 7 for more on this). Insomnia often goes away by itself when the underlying stress that's keeping you up is resolved. You can speed it on its way with the sleep hygiene tips you will learn very soon.

Napping

If you've had a bad night's sleep for whatever reason, you may struggle the next day with drowsiness, irritability, and a strong desire to eat lots of junk food (thanks to alterations in your hunger hormones). Even if you slept well, you still may find yourself fighting off drowsiness about eight hours after you got up. That's when sleep pressure, part of your body's natural ebb and flow of energy across the day, begins to build up. Sleep pressure builds up the longer you stay awake and decreases during sleep. It's at its lowest when you wake up after a night of good-quality sleep, then begins to build again after you wake up.

Around mid-afternoon, many people find that a short nap allows them to get over the energy slump and get through the rest of the day.

The ideal nap is ten to twenty minutes long—long enough to be refreshing but not so long that you enter into deeper sleep and feel groggy when you wake up. The point of the nap is that you'll feel alert and productive afterward. The best napping time is usually about halfway between your wake time and your usual bedtime—generally the time when you feel a late-afternoon energy slump. Experienced nappers (we all know at least one champion napper) can fall asleep quickly almost anywhere, even when the conditions aren't ideal, but if you can, nap in a place that's cool, dark, and quiet.

If you're not a napper or can't fit a nap into your day, a short period of exercise can give you the same recharging effect—circle back to chapter 6, "The Home Gym," for some ideas.

The ABCs to Catching ZZZ's

As the evening wears on, your body starts telling you to get ready for sleep. Sadly, we have a tendency to ignore the message. Instead of gradually slowing down, we keep working, playing a video game, watching TV—pushing ourselves to stay awake that extra hour or two. It's not shocking that when you do finally go to bed, you can't turn your brain off enough to fall asleep.

The best way to get a solid night's sleep is with some tried-and-true basic sleep hygiene. A lot of the tips I give here have some scientific evidence behind them,[14] but some are just common sense. Try them and see what works for you.

- *Have a consistent sleep schedule.* Go to bed at the same time each night and get up at the same time each morning, weekends included. Sleeping in sounds fabulous, but that morning in bed on Saturday may actually result in a net sleep deficit for the following few days.
- *Remove the TV, smartphone, computer, tablet, and video games from the bedroom.* Before you roll your eyes, it is totally possible to do this—take it from me. But if you can't, at least turn them all off. You want to avoid blue light from screens that can affect melatonin production and disrupt your circadian rhythm. Read a printed book instead.
- *Avoid large meals, caffeine, and alcohol starting at least three hours before bedtime.* If you're sensitive to caffeine, cut it out even earlier in the day.
- *Exercise for half an hour a day, preferably outdoors.* Being physically active during the day can help you fall asleep more easily at night. Exposure to daylight helps keep your circadian rhythm and melatonin production on track. Some people like to exercise before bed. This doesn't work for everyone—some get too revved up by the exercise to be able to sleep afterward. In the evening, stick to gentler exercise, like stretching or yoga, that's on the relaxing side.
- *Use relaxation techniques before bed to lower your stress and relax you physically.* The key to getting a good night's sleep is to start relaxing before you get under the covers. Slow your evening pace starting an hour before bedtime—use relaxation techniques to lower your stress and relax you physically. Try using a sleep relaxation app,

and check back to chapter 7, "The Zen Corner," for ways to quiet your mind.

Sweet Dreams

Anxiety can definitely be a sleep killer. If you find yourself stressed before bed, try putting a worry doll under your pillow. Worry dolls come from Guatemalan folklore. You tell the tiny figurines your worries at bedtime (one worry per doll), and then put the dolls under your pillow. When you wake up, the dolls have taken away your worries and replaced them with the wisdom you need to handle the worry. This worked so well for my boys when they were young that I started doing it myself. If the dolls aren't for you, you could also transfer the worry to a journal by briefly jotting down your concerns before you sleep.

Remember how lullabies helped you get to sleep when you were little? Try giving yourself a lullaby with some music. Listening to soft music works by triggering the release of dopamine and decreasing cortisol levels.[15] Classical music is a good choice, but there are also a lot of great apps that will play you to sleep with whatever kind of music you like. The music you select will work best if it's in the 70- to 100-beats-per-minute range—this rhythm mimics your heartbeat.

The Bedroom Renovation

Making changes to your bedroom environment can help make it more conducive to good sleep.

Start with the centerpiece of the bedroom: your bed. More specifically, your mattress. Is it saggy or lumpy? Is it more than five years old? If so, chances are it needs replacing (or at least flipping over), but with what? Should you go with a traditional innerspring? Or memory foam? Or a hybrid?

There's no one-size-fits-all answer here. Check out as many as you can and go with what feels most comfortable to you. Top the mattress with breathable all-cotton sheets, pillowcases, and lightweight or cooling blankets that will keep you from overheating during the night. And invest in some good-quality pillows—firmness is up to individual taste, and yours may be different from your bed partner's, so choose accordingly.

Your window treatments are next. You'll sleep best in a dark room, so you want blinds or curtains that will block light effectively. This is particularly important if you do shift work that means your sleeping hours are also daylight hours. You may need blackout shades or heavy drapes that seal off light gaps. They work well, but they're not always a good fit for your bedroom décor. You can try a light-blocking eye mask instead.

Or you can do what I did during medical school: sleep in your closet. True story. In my last year of medical school I lived in a studio apartment that happened to have a fairly large walk-in closet. Since I prefer total darkness when I sleep and because I knew I would have many night shifts where I would have to catch up on sleep during daylight hours, I decided to put a small mattress on the floor of my closet and use that as my bedroom. Though it didn't do wonders for my dating life, in terms of sound sleep, the closet set-up worked like a charm. The Dark Side definitely has its benefits!

Your body likes to be cooler at night—in fact, one study showed that one of the most important contributors to good-quality sleep is the temperature of the bedroom. That ideal room temperature is around sixty-five degrees (give or take a few) for most people.[16] Turn down the thermostat and use fans or air-conditioning as needed. If you feel like this is a bit on the chilly side, think about wearing socks to bed. So sexy, I know. One study showed that keeping your feet warm with socks in a cooler-temperature room helped people fall asleep faster and stay asleep longer.[17]

HOT TIPS

If you're in the perimenopause or menopause years, hot flashes at night and night sweats may be wrecking your sleep. If you haven't renovated your bedroom to cool it down at night, now is definitely the time. You can also cut down on nighttime hot flashes by avoiding alcohol, caffeine, and spicy foods in the evening (some women with severe hot flashes and night sweats need to avoid alcohol completely). Probiotics have been studied as a way to help reduce hot flashes, night sweats, and other annoying menopause symptoms such as vaginal dryness. Do they help? We know that estrogen and other steroid hormones have crosstalk with your gut bacteria, and one recent study did show a benefit of supplementing with a strain of *Lactobacillus*.[18] We need more research on this. Probiotics can't restore your missing estrogen, but they may help in other ways.

Keeping the noise level down in the bedroom is a real challenge—and I'm not talking about a partner who snores. (You, of course, never snore.) Truly soundproofing your bedroom can turn into a major construction project. Simpler steps include softening surfaces like hardwood floors with sound-absorbing carpeting or rugs. Thick curtains or drapes on the windows help block traffic noise. Seal up gaps around the door—this is surprisingly helpful.

Sometimes the easiest approach to bedroom noise is to fight it with more noise. White noise is noise that contains all audible frequencies, with the energy equally distributed across them. This produces a steady humming sound that masks other noises, like traffic sounds or a TV in another room. A fan or humming air conditioner makes white noise. There's also brown noise, which is a lower frequency than white noise and has a deeper sound, like rumbling thunder, which can help people sleep. Whatever you prefer, you can buy inexpensive tabletop devices that make white noise or other soothing sounds, like rain on a roof or ocean waves.

For waking up in the morning, I like sunrise alarm clocks. They wake you by gradually illuminating the room, imitating the dawn. Instead of being jolted awake by a buzzing alarm in a dark room, you wake up naturally and see "daylight," which helps reset your circadian clock for the day. You can adjust the light intensity and duration to get it exactly right for you.

The *Other* Bedroom Activity

The bedroom is meant for two things only: sleep and sex. You've just learned how to remodel your sleeping habits, so now, in the immortal words of Salt-N-Pepa, let's talk about sex. I bet you weren't expecting this book to give you sex advice. Well, it won't, exactly, but there *is* actually a link between sex and your microbiome—that general contractor really gets around!

You may recall that a happy gut microbiome produces a lot of the neurotransmitter serotonin in your digestive tract. Strange as it might seem, that has an impact on your sex life. Some gut serotonin ends up in your circulation, where it gets carried to the rest of your body and your brain. In your body, serotonin controls and improves the blood flow to your genitals. The better the circulation in this area, the more sexually responsive you are. In the brain, serotonin plays a role in regulating your mood—including your libido. When serotonin levels in the gut, body, or brain are thrown off, so is everything else, and your sex drive sinks. And nothing will sink your sex drive even further than the gas, bloating, and sudden need for a bathroom that an unhappy gut microbiome can cause. A healthy, balanced microbiome is also linked to higher estrogen levels in women and testosterone levels in men.[19]

So get your microbiome balanced through your Gut Renovation and you may feel your sex drive rise. And speaking of rising, some interesting early evidence suggests a link between certain types of bacteria in the gut microbiome and the risk of erectile dysfunction

(ED). It's too soon to recommend probiotics as a treatment for ED, but in the future, those magic blue pills might contain bacteria instead of a drug.[20]

GUT TOOLBOX 

- *Choose wisely.* Pick a mattress that's right for you. Are you a back sleeper? Look for a medium-firm to firm mattress. Side sleeper? Consider a medium-soft to medium-firm mattress. Tummy sleeper? Look for a firm mattress. The Sleep Foundation website (https://www.sleepfoundation.org) has lots of great info to help you choose, but an in-person trial is always the best. Crazy as it sounds, some mattress stores even let you spend the night to decide on the right choice.

- *Neutralize noise.* White noise blocks sound that can keep you up. If you're traveling without your white noise machine, run a fan or the air conditioner, or check out free apps like Sleep Sounds and Relax Melodies. They have a range of soothing sounds that help you relax.

- *Wash your cares away.* A warm bath or shower before bed can be very relaxing mentally and physically. Mentally, you wash away your stress; physically, the warm water soothes and loosens your muscles. It also helps set your body temperature (you actually cool down after a hot bath) to a more optimal sleeping temperature.

- *Smell your way to sweeter dreams.* Spritz your pillowcases with a soothing lavender spray to help send you off to La La Land.

CHAPTER 9

THE NURSERY: HEALTHY GUTS, HEALTHY KIDS

All that restorative sleep and mind-blowing sex you're enjoying in your newly renovated bedroom may lead to an addition to your home—a nursery. Since you now know the pivotal role the microbiome plays in our health, it should come as no surprise that it also is a crucial element of our development from day one, or actually from day zero. Just as a pregnant woman eats for two, her microbiome acts *on* two—her own, and her baby's. That's why this aspect of the Gut Renovation is so important, because the development of a baby's microbiome begins well before they're even born.

If you have kids or plan to, I'm pretty sure you want to learn about how your child's microbiome affects their health. But even if you don't have children and don't plan on it, you were a child once, right? Read on to find out how your childhood might have affected your gut health and therefore your overall well-being. And, more important, learn how to make up for any potential effects of your childhood right now. At the very least, you may learn about yet another way your parents messed you up. But not irrevocably, I promise!

Fertile Ground

Even before a woman becomes a mother, her microbiome is influencing her children-to-be. Just as the gut and the skin have their own microbiomes with their own communities of bacteria, so does the vagina. The vaginal microbiome contains only about three hundred different bacteria species, far fewer than in the gut. The dominant vaginal species are in the *Lactobacillus* family. Metabolites from these bacteria help maintain the vaginal pH. The pH (a measure of acidity/alkalinity) of the vagina is normally slightly acidic, in the range of 3.5 to 4.5—about the acidity of orange juice. This protects it from unfriendly bacteria and yeast, which thrive in a more alkaline environment.

Sperm have their own microbiome, and they like a nice alkaline environment, preferring to swim where the pH is between 7.0 and 8.5. During sex, the pH level in the vagina naturally goes up, making the environment more alkaline. This protects the sperm as they swim their way to the egg and increases the chances of conceiving.

So if something throws off the composition of the vaginal or sperm microbiome, fertility could be impacted. It's also possible that if your vaginal microbiome lacks enough *Lactobacillus* bacteria to keep it in the right pH range, taking oral probiotics may help restore a better balance.[1]

Our general contractor, the gut microbiome, can also play a role in fertility. The most common hormonal cause of infertility in women is polycystic ovarian syndrome (PCOS). Several studies have shown an association between alterations in the gut microbiome and PCOS. Researchers are studying whether ameliorating these microbiome imbalances can help treat PCOS and its associated infertility.[2]

Bun (and Bugs) in the Oven

The amazing changes a woman's body goes through during pregnancy affect all her body's microbiomes, including those in her gut, her mouth, her skin, and her vagina. During pregnancy, her vaginal bacteria shift to include even more *Lactobacillus* species. This may be Mother Nature's way of protecting her and the growing baby from infections that can lead to premature birth.[3]

BABY BLUES

About 10 to 15 percent of women feel depressed and anxious during pregnancy (perinatal depression) and after birth (postpartum depression). In 2017, a randomized, double-blind, placebo-controlled study (the gold standard in research) looked at 423 pregnant women. From the time they were two to three months pregnant until six months after giving birth, half were randomized to take a probiotic supplement containing a bacteria species called *Lactobacillus rhamnosus*. The other half took a placebo; none of the women or the researchers knew which they were taking. At the end of the study, the group taking the probiotic reported significantly lower depression and anxiety than the placebo group. Pretty amazing!

The hormonal changes of pregnancy can change the composition of the gut bacteria and affect the production of gut serotonin, so it's not surprising that the disruptions affect mood and that a probiotic supplement could help by restoring a better balance. Taking a probiotic supplement instead of a drug has a lot of advantages: it's easy, inexpensive, and doesn't harm the baby in the womb or later during breastfeeding.[4]

A baby's microbiome begins developing even before it's fully born. While still in the birth canal during delivery, the baby is exposed to bacteria from the vagina. Immediately after delivery, the baby gets additional bacteria from exposure to the mother's stool—gross but true, and oh so important. Cuddling and kissing give the baby bacteria from the mother's skin and mouth. The baby gets more bacteria directly into the gut when breastfeeding begins—the bacteria come from the skin around the nipples and from the breast milk itself. From these earliest exposures, the baby acquires a gut microbiome that resembles the mother's vaginal microbiota—and remember, the vagina has a strong predominance of *Lactobacillus* bacteria.

But what if the baby skips the vaginal canal route and comes out of the sunroof instead? About 30 percent of babies in the United States are born by cesarean section (C-section) and thus don't get immediately exposed to maternal bacteria from the vagina and stool. Instead, their first exposure is from the environment around them and from contact with their mothers' skin. Research shows that the gut microbiome of C-section babies is definitely different from that of vaginally delivered babies. The C-section babies have a microbiome that resembles the mother's skin—or even the bacteria from the operating room.[5]

The difference between bacterial colonization from the two different modes of delivery is most apparent from birth to three months. By the time the baby is six months old, the differences have pretty much disappeared. But those crucial first few months are the ones that prime the baby's immature immune system to react correctly to threats. If the gut microbiome resembles the mother's during this time, the baby is better equipped to develop a healthy immune system that responds appropriately. If the gut microbiome isn't that close to the mother's own gut microbiome, studies suggest that the baby is more likely to develop allergies that cause wheezing, childhood asthma, allergic rhinoconjunctivitis (hay fever), and atopic dermatitis (eczema). The C-section baby may also be more likely to become obese and develop diabetes later in life. Other health problems that are associated with C-section birth include a great risk of connective tissue disorders, juvenile arthritis, inflammatory bowel disease, immune deficiencies, and leukemia.[6]

Preterm infants are also more prone to develop gut dysbiosis, which puts them at higher risk for a severe intestinal infection called necrotizing enterocolitis, especially if they are in the neonatal ICU.

Studies now show that supplementing these preemies with probiotics can help prevent this potentially fatal complication.[7]

Rock On in Your Rocker

If you're a new mom, you want the renovation of the nursery to include a comfy nursing chair or rocker. Why? Because the best way to help your baby develop a strong gut microbiome, no matter how they were delivered, is by breastfeeding as much and for as long as possible. Breastfeeding appears to counteract the detrimental effects of C-section delivery and helps the baby develop a microbiome that's much more like that of vaginally delivered infants.[8] In large part, that's because breast milk shares some of what's in the mother's microbiome with the baby, and contains special sugars called human milk oligosaccharides (HMOs) that are particularly good for encouraging the growth of beneficial bacteria in the baby's gut.

The right bacteria are important not just for helping the baby digest milk but also for keeping infection away—in fact, for the first few weeks after birth, the baby's immune system is depressed so these bacteria can colonize more effectively. At the same time, breastfeeding passes on the mother's immunoglobulin A (IgA) antibodies, which foster a stronger intestinal barrier and more protective gut bacteria. The remarkable thing is that these benefits from maternal antibodies last into adulthood.[9]

If breastfeeding exclusively isn't an option, or if you need to supplement breastfeeding with formula, don't worry. You can still give your baby's microbiome a great start. While a breastfed baby's gut microbiome gradually comes to be dominated by members of the Bifidobacteria family, formula-fed babies have fewer bifidobacteria and a wider range of other bacteria. The complex natural sugars (HMOs) found in breast milk act as prebiotics to support the growth of bifidobacteria. To mimic this as far as possible, many infant formulas now incorporate HMOs and prebiotics from inulin, a type of soluble fiber.[10]

Probiotic supplements aren't recommended for newborns up to three months, but after that, they too can be added to formula.[11] Very recently, a study looking at supplementing newborns with a probiotic aptly named *B. infantis* for three weeks found that the beneficial effect on the infant's microbiome lasted up to a year.[12]

WEIGHTY INHERITANCE

My overweight patients often tell me it's genetic—obesity runs in their families. But it's not only genetics, it's also bacterial. The gut microbiomes of overweight and obese women have a different mix of bacteria. The obesity-associated microbiome can get passed on to their children through close contact and breastfeeding. A study in Norway showed that there's a strong association with having obesity-associated bacteria in the gut at age two and being overweight at age twelve. Among the children in the study, the makeup of the gut microbiome at age two explained over 50 percent of the variation in later weight.[13]

Dirt Don't Hurt

As a gastroenterologist and the mother of two young boys, I want my kids to get dirty! Why? Because plenty of exposure to bacteria at a young age builds a strong immune system and balanced microbiome. In our hyperclean society, where hand sanitizer is everywhere and kids don't play outdoors very much, many aren't getting enough exposure to a wide range of bacteria from the world around them. Their developing immune systems aren't getting regular workouts and learning how to distinguish between harmful and nonharmful bacteria in the environment. When kids don't get this sort of exposure, they don't develop robust immune systems that will protect them throughout their lives. Instead, their immune system tends to get hyperactive and confused. It starts mistaking harmless substances, like pollen and food proteins, for dangerous invaders and attacking them. In other words, they have allergies.[14]

The concept that dirt is good for kids was first developed in the 1980s, when the rates of childhood asthma, eczema, and food allergies started to skyrocket. The increase coincided with an increasing tendency for kids to stay indoors instead of playing outside and an increasing tendency toward greater cleanliness and greater use of antibacterial chemicals.

For more evidence supporting this hypothesis, we can look at what happens in day care. Babies who go to day care tend to get sick in their first year more often than kids who are cared for at home. They get more colds, ear infections, and gastrointestinal illnesses, which sounds like a nightmare—but as they get older, they get sick less often than the kids who didn't go to day care. The protection lasts until at least age six and probably even longer.[15]

But the hygiene hypothesis has gotten its own renovation recently. Rather than focus on hygiene and exposure to infections, it's now thought that a more important determinant of a child's immune system development is exposure to beneficial bacteria that are less present in the modern environment. This is the microflora hypothesis: the rise of allergic diseases, especially asthma, correlates with the rise of dysbiosis caused by more antibiotic use early in life, an altered gut microbiome, and dietary changes for the worse since the 1980s.[16] Today, about 5.5 million children under age eighteen have asthma—about 7.5 percent of the total population in that age group.

Though we know that gut dysbiosis is associated with an increased risk for asthma, what's still unclear is if fixing the underlying dysbiosis with probiotic supplements can help prevent and treat asthma. The research to support this just isn't solid enough yet.[17] But the link between the infant microbiome and asthma is clear. A Canadian study categorizing the microbiomes of infants showed that if they had an abundance of four specific families of bacteria (*Lachnospira*, *Veillonella*, *Faecalibacterium*, and *Rothia*) during the first one hundred days of life, then their risk of developing asthma later on was significantly reduced.[18]

The evidence is even stronger for using probiotics to prevent and treat another common childhood ailment, atopic dermatitis (AD), also called eczema. Eczema affects as many as 20 percent of all kids, causing dry, itchy skin and red rashes. It usually starts in infancy and often goes away by the teenage years. Eczema is often the first step in the "atopic march," a series of allergic disorders that can last a lifetime. First comes food allergies, then hay fever, and then wheezing and asthma. Kids with eczema are also at greater risk of developing inflammatory diseases such as ulcerative colitis later in life, and they're more likely to have ADHD. Eczema tends to run in families, and there are some specific genes that are associated with it. Researchers have explored giving a pregnant woman probiotics if the baby is likely to inherit a tendency to eczema. The idea is that the probiotics would calm the immune system and keep it from overreacting and causing allergic symptoms. There's enough positive evidence for probiotics in preventing eczema in at-risk babies that the World Allergy Organization suggests using probiotics in pregnant and nursing women and in bottle-fed infants at high risk of developing allergies.[19]

Kids who grow up on farms around a lot of animals have a lower risk of allergies, probably because they're exposed to animal allergens from infancy. Amish kids, for instance, grow up on old-fashioned farms without mechanical equipment. They're constantly exposed to cows, horses, and other farm animals from birth—and their asthma rate is low, even compared to kids who grow up on today's mechanized farms. Something similar happens to kids who grow up with cats and dogs. The more cats and dogs in the household when a kid is a baby, the lower the chances that they'll develop eczema, hay fever, or asthma later on in childhood. The effect is dose dependent: the more animals, the lower the risk.[20] As part of your Gut Renovation, add a dog bed and a cat scratching post!

IN THE DOGHOUSE

Let's not forget our fur babies! Just like you, your dog's gut microbiome can get out of balance and leaky through eating the wrong foods (too much people food, usually), from food allergies, and from being sick with a virus or food poisoning. Probiotics can help calm a dog's digestive system, reduce allergic reactions, and reduce the emission of potent doggy gas. They're also really helpful for improving dog diarrhea caused by stress colitis (from being at the vet or boarded, for example), from sudden changes to the usual diet (eating something found in the woods or getting into the cookie jar, for example), and from antibiotics.[21] Prebiotics can also help your dog's gut health. For example, to help with diarrhea symptoms, add a few spoons of plain canned pumpkin (not the kind with spices for pumpkin pie) to the dog's regular meal. The soluble prebiotic fiber helps dogs the same way it helps people—and it works for constipation, too.

In terms of probiotics, rather than giving your dog the same probiotics you take, look for dog-specific brands—or ask your vet. You can also try adding small amounts of unsweetened plain live-culture yogurt to your dog's regular food.

Autism and the Microbiome

Autism spectrum disorder (ASD) is another condition that's rising in incidence. ASD is a developmental disorder that may also be triggered by gut dysbiosis. In addition to the behavioral and language development differences that start to appear in early childhood, young kids with ASD often (as many as 70 percent) have gastrointestinal symptoms such as diarrhea, abdominal pain, constipation, and reflux. One study showed that kids with ASD are four times as likely as other children to have gut problems.[22] Another recent study showed that both kids and adults with ASD are more likely to have inflammatory bowel disease.[23] Kids with ASD show imbalances in their gut bacteria, with higher levels than usual for some harmful bacteria families—classic dysbiosis.[24] In fact, a recent study that looked at seventy-two families and compared the microbiomes of autistic children to their nonautistic siblings found distinct differences. This is quite striking, because the siblings have similar genetics and live in the same home, so they probably have similar environmental exposures and diets.[25]

Probiotics for Kids

Probiotics can be a real help for treating some common childhood illnesses. They're not a cure, but they can help relieve symptoms and may even shorten the duration of the illness.

Kids with acute gastroenteritis, otherwise known as a stomach bug, can benefit from probiotic supplements added to the usual treatment of a few days of rehydration and rest. Studies show that some well-understood probiotic bacteria strains, such as *Lactobacillus reuteri* and *Saccharomyces boulardii*, can help shorten the duration of watery diarrhea by a day and reduce the risk of spreading the infection by reducing stool volume. Shortening the illness by only a day may not sound like much, but as a parent I can tell you, it can make life easier for the whole household, not to mention that it could make the difference between being able to stay home and needing to be hospitalized for dehydration.

Is your family soundtrack the sound of coughing from a cold? Up to age two, kids average eight to ten colds a year. Sharing germs in kindergarten can raise the number to twelve a year. Giving your child probiotics twice a day during cold season (October through

April) may help reduce how often your child gets a cold and may also shorten its duration.[26] They may be particularly helpful for reducing nasal congestion and runny nose symptoms.[27]

What you don't want to do is give your child antibiotics for every cold or cough or even every ear infection. This isn't always easy. In fact, one of the most difficult decisions a parent faces is whether to give a child antibiotics for an illness. Trust me, I can relate. Nobody likes to see their child sick or in pain. When my younger son was five years old, he had a string of painful ear infections, and seeing him suffer through them made me wish I could just knock them out with some heavy-duty antibiotics. But most ear infections are caused by viruses, which aren't killed by antibiotics. Viral infections only rarely lead to bacterial infections, so preventive antibiotics aren't needed. Knowing this and knowing how damaging antibiotics can be (and following the guidance of our pediatric ENT), I resisted that urge and allowed the infections to clear on their own.

Your pediatrician can help you decide if the infection is viral or bacterial and help you decide if an antibiotic is needed. Antibiotic use early in life, especially before age two, can cause a long-lasting shift in the makeup of your child's microbiome that can affect their future health. The more antibiotic prescriptions a child gets, the greater the risk—and the greater the risk of having a combination of conditions.[28] The shift is associated with an increased risk of asthma, atopic dermatitis and other allergies, obesity, inflammatory bowel disease, and an increased risk of resistance to some antibiotics.[29]

Sometimes your kid does get really sick with something that needs an antibiotic to fix, despite the risks, and the pediatrician recommends you move ahead with one. If an antibiotic is truly needed for a bacterial infection, be on the lookout for side effects. About one in ten kids will have a side effect, such as vomiting or diarrhea, bad enough for an emergency room visit.[30]

In the gut, antibiotics kill both beneficial and harmful bacteria. That can cause diarrhea, the most common side effect in babies and kids. You can counteract the risk of diarrhea by giving your child probiotics to restore the good bacteria—in one meta-analysis, taking probiotics with antibiotics reduced the risk by more than 50 percent.[31] Be careful with this, however. If you give the probiotics and the antibiotic at the same time, the antibiotic will kill the probiotics before they have a chance to help. Separate the doses by a couple of hours. Look for a probiotic product that contains many different species, not just one or a few.

Little Tummies, Big Consequences

As a parent, you want to make sure your child's diet is optimal for building a strong gut microbiome. During the first three years of life, what your child eats will set the broad parameters of their microbiome for life, with the potential to affect their short-term and long-term health. No pressure, right?

As a baby transitions from a milk-based diet to eating more solid foods, their microbiome changes. A different array of bacteria begins to take over, and some of the babyhood bacteria decrease in numbers. The difference can be detected by the time the baby is age two. By age three, the child's microbiome is starting to resemble that of an adult. By age four, the impact of diet on the gut microbiome is very clear. Kids that eat a typical American diet—high in refined carbohydrates, juice and sweetened beverages, kid's meals, and lots of snacks and sweets—have a clearly different bacterial population than those who eat a diet with less sugar and more whole foods. The more refined grains and sugar in the diet of a young child, the less diverse the microbiome. The standard low-fiber American diet may influence a child's gut microbiome in ways that increase the risk of allergies, inflammatory bowel disease, and metabolic disorders such as obesity and diabetes later in life.[32]

Just as most adults don't get enough dietary fiber, most kids don't either. (Check back to chapter 4, "The Bathroom," for why fiber is so important.) Kids with a better diet that has more fiber also have more bacterial diversity in their gut microbiome and have more of the bacteria that increase calcium absorption—important for growing strong bones. They may also produce more short-chain fatty acids, which are important for overall gut health and for preventing colon cancer down the road.[33]

The good news is you already know how to feed a healthy microbiome: the same dos and don'ts you learned in chapter 3, "The Kitchen," apply to children as well. They're actually even more important to follow, because with childhood nutrition you have a real opportunity to prevent a host of lifelong illnesses and foster optimal long-term health. So gather the whole family together for some gut-healthy meals (recipes coming at the back of the book!) and enjoy! Luckily, one probiotic food that most kids enjoy and are willing to eat daily is yogurt. Just make sure you choose one without excessive sugar and which clearly contains live and active cultures. If they don't like to eat yogurt, try a fermented milk drink like the Japanese Yakult, a fan favorite in my kids' school.

Exercise and sleep are also equally if not more important for kids than adults, as is spending time away from devices. These were all particularly challenging during the pandemic and some bad habits were formed. One of the best ways to encourage your kids to engage in healthy practices is to lead by example. Yelling at your kids to get off their iPads when you're glued to your phone doesn't work—especially if they're old enough to know the word *hypocrite* (trust me, I've been there). Planning family time when you're all device-free and being physically active—this could be as simple as a fifteen-minute walk outside—is a great way to remind everyone (including yourself) that health is a priority.

GUT TOOLBOX 

- *Cuddle up.* Have lots of skin-to-skin contact with your baby to pass on your beneficial bacteria. Cuddle, kiss, and breast-feed, especially if your baby was born by C-section.

- *Shop savvy.* Kids' microbiomes are just as easily damaged by antibiotics as ours are. Shop for antibiotic-free dairy and animal products to minimize your children's exposure.

- *Intervene early.* If allergies run in your family, talk to your pediatrician about the right age to start giving your baby probiotics. They can help prevent childhood allergies such as atopic dermatitis (eczema) and wheezing.

- *Make fiber fun.* Make eating gut-healthy food fun for little ones with a fiber point system. Assign different fiber points to various foods and allow them a treat if they reach ten points.

CHAPTER 10

THE LAUNDRY ROOM: DETOXING YOUR HOME

The final space to tackle is the laundry room. As you know if you've remodeled your home, *every* renovation involves a thorough clean-out at the end, once the other pieces are in place. The same goes for your Gut Renovation. After all, you wouldn't leave piles of dirty laundry all over the place in a spotless new home, right? Of course not!

Even with the best habits in the world, you need to make sure your environment is conducive to your wellness. In this chapter, you'll learn to clear out harmful toxins that can sabotage your gut health—and you may be surprised to find that those toxins are lurking in your home and everyday products.

Yes, You Can Be Too Clean

If you skipped chapter 9, "The Nursery," because you don't have kids, go back and read it now. That's where I explain how being too clean can actually be harmful to your immune system and your gut microbiome. That doesn't mean you need to live in a dirty house and wear smelly clothing. It simply means you accept that bacteria are all around us, on us, and in us. Instead of constantly cleaning with harsh chemicals to try to annihilate them, choose to clean strategically, using products that do the job well but don't disrupt your internal and external environment.

The first cleaning products to replace in your laundry room renovation are antibacterial soaps and cleaning products. The ads for these products hint that they keep your family safe by killing germs, but in fact, there's no evidence that they work any better than plain soap and water for keeping your skin and home clean. There's plenty of evidence that shows just the opposite, however—soaps and cleaning products with antibacterial chemicals actually contribute to the very serious problem of antibiotic resistance. (I'll explain why the rise of superbugs—bacteria that are highly resistant or even immune to antibiotics—is so concerning later in this chapter.)

In 2017, the FDA effectively banned the two most widely used antibacterial ingredients, triclosan and triclocarban, along with some fifteen other antibacterial products. Animal studies showed that triclosan causes rapid changes in the diversity and composition of gut bacteria, and the manufacturers couldn't provide clinical evidence that their products worked better than regular soap without antibacterial ingredients. At the same time, the FDA also pointed out another problem: we don't know how safe antibacterial products are in the long run. Manufacturers have dropped the antibacterial claims and stopped using these chemicals in consumer products, but they're still allowed in medical settings like hospitals and clinics.

The ban went into effect in 2018, so it's possible you still have some of these products in your home. If you see the words triclosan or triclocarban on the packaging or label, toss the product.

Keep the Hand Sanitizer

While I want you to get rid of all those old antibacterial soaps and cleansers, don't throw out the alcohol-based hand sanitizer! Moving forward from COVID, even though we know surface transmission is by no means the biggest risk, we have a new appreciation for hand sanitizer. Many of us are realizing that we were actually bringing a *lot* of germs and toxins into our home, when it was preventable. These products are safe and effective when soap and water aren't available. If it's at all possible, though, wash your hands instead.

Cleaning Products and Your Microbiome

If a cleaning product kills bacteria in your environment, can it also harm the bacteria of your inner environment—your gut microbiome? Yes, which is another good reason to use safer alternatives and be a little less fanatical about cleaning everything on and around you. Anything you put on your skin or breathe in can be absorbed into your body and from there can impact your gut microbiome. Triclosan, for instance, is known to affect the gut microbiome in ways that cause inflammation.[1] Recent studies also suggest that infants exposed to household disinfectants have gut microbiome changes that can lead to the child being overweight or obese by age three.[2] We also know that exposure to household chemicals is linked to wheezing and asthma in babies.[3]

If you have to use heavy-duty detergents and disinfectants—to clean a toilet, for example—wear gloves and be sure the area is well ventilated. Keep kids and pets out of the area until all surfaces are dry and any odors have dissipated.

SAFER ALTERNATIVES

Any plain liquid or bar soap works well to remove grime and bacteria from your hands and the rest of you. For heavier household cleaning, look for products that have the EPA's Safer Choice label. All the ingredients in these products meet the EPA's stringent safety requirements, so they're safer for you and for the environment. You can find a list of certified products on the Safer Choice website at https://www.epa.gov/saferchoice. The nonprofit Environmental Working Group (https://www.ewg.org) also has lots of great information about safe cleaning products. Do-it-yourself safe cleaning products are inexpensive and easy to make. The internet is full of handy formulas using nontoxic ingredients like vinegar and citrus oil. Experiment and find formulas that work well for you. Be careful with creating your own formulas, though. Combining bleach and ammonia, for example, creates toxic chloramine gas that can kill you. And mixing bleach with hydrogen peroxide releases oxygen so rapidly that it can cause an explosion! Don't mix bleach with any other cleaning products.

Preventing Antibiotic Resistance

Every year in the United States, about 2.8 million people get an antibiotic-resistant infection—and more than thirty-five thousand of them die. Your medicine cabinet could be contributing to the problem.

When you take an antibiotic to treat an infection, the drug kills not only the harmful bacteria that cause the infection, but also some of the bacteria in your gut—including the ones you need to keep your microbiome in balance. That's why diarrhea is a common side effect of antibiotics. A related problem is the way inappropriate antibiotic use drives the growth of antibiotic-resistant bacteria. When a doctor prescribes an antibiotic, the targeted dangerous bacteria are destroyed. A few, however, may have natural variations that let them survive the attack. When these hardy survivors multiply, they pass on their ability to resist the antibiotic, creating a new population of bacteria that are harder to kill. Eventually, some of them become superbugs—meaning they can't be killed at all.

The more we use antibiotic drugs and antibacterial cleansers, the more we create resistant bacteria. There are times when antibiotics are needed, and the benefits outweigh the risks of side effects and creating resistance. But when antibiotics aren't needed, you get all the risks with none of the benefits.

In my practice, of course I prescribe antibiotics when they're needed. But I also spend a lot of time explaining to my patients why I'm not prescribing an antibiotic for them, because I have an obligation to be a good steward of these drugs. If I think a patient has a viral infection, for example, an antibiotic won't help, because these drugs only kill bacteria—and it might hurt by damaging their microbiome for no good reason and adding to the problem of drug-resistant bacteria. If they do need an antibiotic, I encourage them to increase their intake of probiotic-rich foods and in some cases to take a probiotic supplement as well.

During the COVID-19 pandemic, a lot of doctors prescribed antibiotics for their patients, even though these drugs don't kill viruses. They were concerned about secondary bacterial infections in patients who were weakened by the viral infection, so they prescribed the antibiotics as a preventive measure. In the first half of 2020, when the pandemic was going strong, just over half of all patients hospitalized for COVID got antibiotics, even though most of them didn't have a bacterial infection.[4] At that point, we didn't have a good understanding of the best ways to treat COVID-19, so the overprescribing is understandable. As we've gained experience with COVID-19, overprescribing is becoming less of a concern, but it's still happening and could create even more antibiotic-resistant superbugs in the future.

To do your part in preventing antibiotic resistance, bear in mind that antibiotics don't usually help if you have a sore throat, a cold, or the flu. Everybody has unique health needs, however, so discuss your need to take an antibiotic with your doctor. Take them exactly as your doctor tells you and take all of them—don't stop after a few pills because you're feeling better. Don't share your antibiotics with others, and don't take antibiotics that were prescribed for someone else.

FECAL TRANSPLANTS

Overusing antibiotics damages your gut microbiome the same way a forest fire burns down the woods—good and bad bacteria get killed alike. When the gut microbiome is damaged, you can't defend yourself as well against infections (even though you took the antibiotic to get rid of an infection). That opens the door to infection from a very nasty gut bacterium called *Clostridioides difficile*, or *C. diff* for short. It causes severe diarrhea that's very hard to treat. Every year in the United States, *C. diff* causes more than three million infections and about thirteen thousand deaths.

One treatment for *C. diff* is a fecal transplant, also called fecal microbiota transplantation (FMT). A fecal transplant transfers feces from a healthy donor into someone else, with the goal of restoring a healthy balance of bacteria in the gut of the recipient. How exactly do they do this, you may wonder? Several methods: One involves delivering the stool through either an enema or during a colonoscopy. Stool capsules that you swallow to get the donor feces are also used. As yucky things go, a fecal transplant is way up there. They used to also be way out there, seen as a fringe idea. No longer—fecal transplants are on their way to being mainstream medicine.

Environmental Chemicals

Environmental chemicals such as car exhaust, paint fumes, plastic microparticles, insecticides, and thousands of others are all around us all the time. They're an inescapable part of modern life, and because they surround us, they end up having an impact on your microbiome. Exactly what the impact is depends on the chemical and your exposure.

In general, however, environmental chemicals interact with your gut microbiome in several different ways. Your gut bacteria actually like to eat some chemicals, but that can change the metabolites they produce. Other chemicals get absorbed into the bloodstream through the walls of the gastrointestinal tract and end up being carried to your liver for detoxification; that process modifies the chemicals and sends them back through your gut for excretion—where your bacteria act on them again and may form new toxic metabolites. The chemicals can also alter the balance of bacteria types in the gut, possibly causing a leaky gut and dysbiosis. And finally, environmental chemicals can change the metabolic activity of the gut bacteria and throw off the balance.

Scary, right? You can't avoid every environmental chemical out there, but you can try to reduce your exposure to some of the most common with some fairly easy steps.[5]

Semivolatile organic compounds, or SVOCs, are very widely used in home products. Since the 1970s, flame retardants, a type of SVOC, have been added to many, many household items, especially upholstered furniture, electronics, mattress foam, and polystyrene building foam. Plasticizers (substances added to materials to make them softer and more flexible) and pesticides are other examples of almost inescapable SVOCs. You're exposed to these chemicals all the time, mostly through indoor dust as the SVOCs leach out, wear off, or get applied. SVOCs are associated with a lot of health issues, such as endocrine disruption and neurological damage. Little kids are the most vulnerable, because they crawl around on the floor with the dust and then put their fingers in their mouths. Aside from the other damage these chemicals can cause, they can also disrupt your microbiome, altering the types and numbers of bacteria.[6]

You can reduce your exposure to flame retardants by keeping dust levels down through wet mopping and using a vacuum cleaner with a HEPA filter. When you buy new products and furniture, look for cotton, polyester, or wool in the cushioning instead of polyurethane foam. Avoid using chemical pesticides—check out integrated pest management (IPM) techniques instead. IPM uses environmentally friendly methods to control household pests like rodents and insects with no or minimal use of poisons or other dangerous chemicals.

Indoor Air Quality

We spend about 90 percent of our lives indoors, so it's important to pay attention to the indoor air quality of your home and your work environment (which for many of us are the same since 2020!). Indoor air pollution can come from a wide range of sources. The semi-volatile organic compounds from household cleaning products are one common source. Many other sources release volatile organic compounds (VOCs) that are just as bad for you. VOCs come from paints, paint strippers, and other solvents, aerosol sprays, air fresheners, dry-cleaned clothing, pesticides, crafts materials such as glue and adhesives, and office equipment such as copiers and printers. On top of the SVOCs and VOCs, indoor air pollution can come from mold, building materials, carpets, combustion sources such as a gas stove or fireplace, and personal-care products.

Exposure to indoor and outdoor air pollutants changes the composition of your gut microbiome, making it less diverse and potentially shifting the balance toward harmful bacteria.[7]

You can take steps to reduce your exposure to indoor air pollutants, such as choosing low- or zero-VOC paints, using a green dry cleaner, and tracking down and eliminating mold. The best way to reduce indoor air pollution, however, is by better ventilation. Since that's also the advice for reducing transmission of COVID-19, now's the time to make sure your Gut Renovation includes window treatments that let air flow freely and can handle window fans. It might also be time to invest in some portable air purifiers for times when you don't want to bring in outside air, like when it's really cold out or when the outdoor air has a lot of pollution or allergens. For an air purifier to be effective, it needs to be the right size for the room and have a HEPA filter.

Buy Organic

You can reduce chemical exposure in the kitchen by buying organic produce.

Industrialized farming uses a lot of pesticides, insecticides, fungicides, weed killers, and chemical fertilizers. You inevitably end up consuming some of these chemicals. We're still learning what the cumulative effect is on your digestive system, but chemicals that kill fungi or insects on plants likely also kill bacteria in your gut.[8]

To avoid the problem, look for organically grown food. The USDA organic label means the produce was grown on soil that had no prohibited substances, like synthetic fertilizers and pesticides, applied for at least three years prior to harvest. The USDA standard is so rigorous that many small growers, the kind who sell at your local farm market, can't meet it, even though they use organic methods and don't apply chemicals. It's safe to buy from them; plus, buying local produce supports farmers and maintains open space in your community. And nothing could be fresher than produce picked that day.

Organically grown produce can be a bit more expensive than standard supermarket prices. If you can't afford or can't find organic versions of some foods, go ahead and buy the conventionally grown produce instead. The Environmental Working Group puts out a Clean Fifteen list of produce that's safe even when it's not organic. The list varies a bit from year to year, but generally includes asparagus, avocados, broccoli, cabbage, cantaloupe, cauliflower, sweet corn, eggplant, honeydew melon, kiwi, mushrooms, onions, papaya, frozen peas, and pineapple. The EWG also has a Dirty Dozen list of fruits and veggies to buy only in organic form—these are the foods that are most heavily sprayed with pesticides. The list includes apples, bell and hot peppers, celery, cherries, grapes, kale/collard/mustard greens, nectarines, peaches, pears, spinach, strawberries, and tomatoes.

Nonprescription Drugs

The shelves of any pharmacy are crammed with nonprescription drugs (also called over-the-counter drugs, or OTCs). OTC antacids and digestive aids are great for dealing with minor issues like occasional heartburn, gassiness, or mild diarrhea. By definition, the FDA says these drugs are so safe and effective that you don't need a doctor to prescribe them. You still need to be very careful about taking them—some of the most widely used nonprescription drugs can do some serious damage to your digestive tract and throw your microbiome out of balance. If you find yourself taking OTC antacids all the time, speak to your doctor, as you may have a bigger problem that needs to be addressed.

Top of the list of over-the-counter drugs are nonsteroidal anti-inflammatory drugs (NSAIDs), including aspirin, naproxen (Aleve), and ibuprofen (Advil, Motrin). These drugs relieve pain, reduce swelling, and bring down fevers. Acetaminophen (Tylenol) isn't an NSAID because it works in a different way, but this drug is also OTC and is very widely used as a pain reliever and fever reducer.

Taking an NSAID now and then for a sore muscle or a headache is fine. But as I see in my office every day, taking NSAIDs several times a day, every day, can cause some serious gut problems. These drugs can irritate the stomach and gut lining and damage the protective mucus layers, causing gastritis and even ulcers. They can cause abdominal pain, severe stomach or intestinal bleeding, and increased intestinal permeability, aka leaky gut, and they can alter your gut bacteria.[9]

Acetaminophen, on the other hand, doesn't cause stomach or intestinal irritation and doesn't seem to affect your gut microbiome. It's a good alternative to NSAIDs for pain and fever, although it doesn't help inflammation. Use this drug cautiously, though, because in large doses it can damage your liver—keep your daily dose to under 3,000 mg and don't drink alcohol. You may be getting more than you think because acetaminophen is added to a lot of OTC formulas like cold and allergy remedies. Also, be very careful about keeping it away from kids and pets. Even a small piece of a single acetaminophen tablet can kill a cat.

It's Five o'Clock Somewhere

The COVID pandemic was stressful (understatement of the year). Between lockdown, working from home, dealing with homeschooling, politics, and of course the very real health dangers, it's not a surprise that some of us ended up drinking more than usual as a way to cope. For context, according to the National Institute on Alcohol Abuse and Alcoholism, moderate alcohol use for healthy adults generally means up to only one drink a day for women and up to two drinks a day for men. But a nationwide survey done by the RAND Corporation in the spring of 2020 during the lockdown period showed that compared to the same time the previous year, women were drinking 14 percent more and men were drinking 17 percent more. The survey also showed a worrisome 41 percent increase in heavy drinking (eight drinks or more a week).[10]

As life returns to something like normal, now is a good time to consider cleaning out the liquor cabinet—cutting back to your usual alcohol consumption level or less, or even giving up on alcohol completely. The stress management techniques I talked about back in chapter 7, "The Zen Corner," can be helpful tools for scaling down the booze intake.

And if you need more incentive, I'm here to tell you that alcohol has really detrimental effects on your gut microbiome. Think of it this way: For years, you've been using alcohol-based sanitizer to kill microbes on your hands. The alcohol you drink does the same thing to the bacteria in your gut—and it's very damaging. Alcohol alters the balance of bacteria in your gut, shifting it to different dominant bacteria families. If you drink a lot, you're much more likely to have dysbiosis and SIBO caused by disruptions to your gut bacteria. Dysbiosis from alcohol and inflammation from gut wall permeability seem to play important roles in developing alcohol-related liver disease. On the other hand, the polyphenols in red wine can shift the gut microbiome in a more positive direction—but only if red wine is drunk in moderation. More than a glass a day and the benefit is lost. Some of the damage to the gut microbiome from alcohol will improve just by stopping or cutting back on drinking. Prebiotic and probiotic supplements can help speed up the repair process.[11]

Alcohol impacts your microbiome in less obvious ways as well. It disrupts your sleep, makes you hungry for sugary and salty processed foods, and makes you less able to resist them. As you know from chapter 8, "The Bedroom," your microbiome needs its rest just as the rest of your body does. And you know from just about every chapter of this book how bad processed foods—and alcohol counts as one!—are for your gut.

Kindly Refrain from Smoking

One of the most important Gut Renovations you can possibly make is a simple one: throw away the cigarettes. Aside from all the other ways smoking can kill you, it can also seriously harm your whole digestive system. There's the increased risk of cancer everywhere from your mouth to your anus—smoking is a major risk factor for oral cancer, esophageal cancer, and colon cancer, the second-leading cause of cancer death. Then there's the link between smoking and heartburn and GERD, and the link between smoking and stomach ulcers. And then there's the damage smoking does to your gut microbiome. When the gut bacteria of smokers are compared to the gut bacteria of nonsmokers, you see an imbalance in the smoker bacteria profile; this suggests that for conditions such as inflammatory bowel disease (which are related to alterations of the gut microbiome), quitting smoking should be part of the treatment. When smokers quit, their gut bacteria gradually rebalance back to the nonsmoker profile.[12]

And there you have it—all the tools you need for your Gut Renovation. Now it's time to see how to use them in your daily life. The Gut Reno program awaits you in the next and final chapter, "The Living Room."

GUT TOOLBOX 

- *Bug spray.* Want to be really cutting-edge? Invest in a probiotic air purifier. This innovative product actually sprays good bacteria into the air of your home (or office) where, according to the manufacturer, the bacteria consume the pet dander, dust mites, pollen, and other allergens that would normally trigger allergies in your indoor environment.

- *Clean consciously.* Make your own all-purpose cleaner: Mix equal parts water and distilled white vinegar in a spray bottle. Add ten drops of a fragrant essential oil like lemon, peppermint, or tea tree. Shake well and spray away the grime safely.

- *Remodel happy hour.* If you and your coworkers are trying to cut down on drinking, then make your work events less alcohol dependent and more health oriented. Happy hour could transform into Frisbee in the Park Hour or a group Healthy Cooking Class. The best part? No hangovers at work the next day!

- *Shop local, think global.* Shopping at your local farmers market not only allows you to buy locally grown organic produce—adding lots of local color to your plate—it's also better for the environment, because it cuts back on long-distance driving for food delivery.

CHAPTER 11

THE LIVING ROOM: THE GUT RENO PROGRAM

Now it's time to put everything you just learned into practice to truly transform your gut and your entire body into your dream home. In the Gut Reno program you'll find gut-friendly options to optimize your health. But of course we don't all have the same tastes. You're the *interior* designer here (literally), so you can choose whatever options suit you best. This is important because, as I tell my patients, consistency is the key to sticking to a healthy habit, and you won't be consistent with something that feels like a chore.

Something I do want to stress is that it's not feasible to do everything at once—and that's okay! It's not like you'd renovate all the rooms in the house simultaneously. Now, though, you know how all the pieces fit together, and you can modify the plan to best suit your individual needs.

What I love about the Gut Reno program is that it's realistic and doable. The scrumptious recipes in particular (you'll find them in the back of the book) are simple and quick, mainly because I'm an impatient chef with two hungry boys and not a lot of time. The exercises are easy to incorporate into a busy schedule too, even if you have only ten minutes to spare. So explore, experiment, and discover what works for you and your lifestyle. And as I said before, don't feel you have to tackle everything at once. Just commit to breaking ground and start your renovation in whichever room you choose first. There's definitely something for everyone in this plan. Just go with your gut!

Your Gut Reno Eating Plan

When you renovated your kitchen, you emptied your cupboards of a lot of processed and convenience foods and sugary and salty snack foods. Now it's time to replace them with better choices that are high in fiber, probiotics, and nutrients, low in added sugar and salt, and full of natural flavor. The foods I list below are great options to optimize your microbiome. They're easy to find in any supermarket, and they make tasty substitutes for processed foods. For instance, swapping out the regular pasta for whole-wheat pasta or chickpea pasta, or white rice for brown rice, is really easy. You may have to adjust the cooking time a bit, but otherwise, you use these staples in exactly the same way—while getting the benefit of more fiber and amazing flavor.

Foods for Fiber

As I explained back in chapter 4, "The Bathroom," eating more fiber is probably the single most beneficial thing you can do for your gut. Plant foods are by definition rich in soluble and insoluble fiber. You want to be eating at least five servings a day of any combination of fruits and vegetables, along with other good sources of fiber such as whole grains, beans, nuts, and seeds. I really encourage you to go beyond five servings and aim for seven or eight a day (but remember, build up to extra daily servings *gradually*!). Your microbiome will thank you by making your digestive processes move along smoothly.

And remember, a serving is only half a cup, or about 2.5 ounces—roughly the amount you could hold in the palm of your hand. The exception is green leafy veggies you're eating raw, like lettuce or spinach. In that form, a serving is 1 cup, or roughly the size of your fist. So, a serving is only two cooked broccoli spears, or one medium-sized carrot, or just half a big apple. Another way to look at it is to aim for 1 cup of fruit and 1½ cups of veggies each day. If you look at your measuring cups, you'll see that in total, that amount isn't really a lot to eat. You can easily manage to add the extra fiber servings, especially if you use them to replace servings of foods that aren't that great for you.

The lists below are of natural foods that are especially high in fiber per serving, but pretty much all fruits and vegetables have some fiber—and some fiber is better than no fiber.

Fruits

Apple
Banana
Blackberries
Blueberries
Fresh coconut
Mango
Orange
Passion fruit
Pear
Raspberries
Strawberries

Grains, Seeds, and Nuts 

Almonds
Barley
Brown rice
Buckwheat
Chia seeds
Oatmeal
Pistachios
Pumpkin seeds
Quinoa
Sunflower seeds
Walnuts

Vegetables 

Artichokes
Avocados
Beets
Broccoli
Brussels sprouts
Carrots
Cauliflower
Pumpkin
Squash
Sweet potatoes

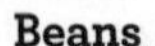

Beans 

Chickpeas	Lentils
Kidney beans	Split peas

EAT LIKE A PRO

To keep your microbiome balanced and healthy, you should aim for one to two probiotic and prebiotic foods every day. The probiotic foods give you extra beneficial bacteria, and the prebiotic foods support their growth.

PROBIOTIC-RICH FOODS

Milk-based probiotic foods include live- and active-cultures yogurt (make sure it says that on the label) and kefir. In the kitchen, you can swap out sour cream, heavy cream, and milk for plain yogurt (which is also a great substitute for mayo and eggs).

Cheeses that contain probiotics include cottage cheese, mozzarella, cheddar, Edam, Gouda, Gruyere, Parmesan, provolone, and Swiss. Sour cream is made by fermenting cream, but the product is then pasteurized, which kills the bacteria. Some brands add live bacteria back in, however—check the label.

Fermented plant foods that contain probiotics include sauerkraut, pickles (but not the kind made with vinegar), kimchi, miso, tempeh, and natto. Kombucha, a very mildly alcoholic drink made from fermented black or green tea, is said to contain probiotic bacteria, but many commercial products don't contain large amounts of live beneficial bacteria, and there is little evidence to support the many health claims associated with this drink.

PREBIOTIC FOODS

As you'll see from the list, you have a lot of choices here, so it's easy to aim for at least one daily serving of a prebiotic food. I like a handful of almonds mixed with a handful of raisins—two prebiotic servings in one snack.

- Acai berries
- Alliums (garlic and onion family)
- Almonds
- Apples
- Artichokes
- Asparagus
- Bananas
- Barley
- Beans, especially white beans, chickpeas, and lentils
- Chicory
- Dandelion leaves
- Flaxseeds
- Green plantains
- Honey
- Jerusalem artichokes
- Mushrooms
- Oatmeal
- Raisins
- Seaweed
- Yucca (cassava)

Some foods that don't have any fiber still have prebiotic benefits because of their natural sugars or polyphenols (a type of phytonutrient). Leading this list is red wine. Others include honey, maple syrup, and dark chocolate.

Phytonutrients

As we learned back in chapter 3, "The Kitchen," *phytonutrients* is a catch-all term for the many natural chemicals found in plant foods. The plant world contains *hundreds* of these. They all act as powerful antioxidants that protect your cells from damage and help them repair any damage that does occur, and are important antiaging nutrients. Eating a diet with plenty of phytonutrient-rich foods is probably the best antiaging dietary advice I can give.

It's simple: just eat the rainbow. The more colorful a plant, or the darker the green, or the more pungent the flavor (think garlic), the more phytonutrients it probably contains. Use the following list as a guide.

PHYTONUTRIENT FOODS

- *Red, orange, and yellow vegetables and fruits:* apples, beets, berries, carrots, citrus fruits, mangos, melons, peaches, peppers, squash, sweet potatoes, tomatoes
- *Dark green leafy vegetables:* arugula, chicory, bok choy, broccoli, collards, kale, dark lettuces, spinach, Swiss chard
- *Alliums (onion family):* chives, garlic, leeks, onions, scallions
- *Whole grains:* barley, brown rice, buckwheat, oats, quinoa, wild rice, whole-wheat bread, whole-wheat pasta, and whole-grain cereals
- *Nuts and seeds:* almonds, flaxseeds, sunflower seeds, walnuts
- *Beans:* all beans, lentils, peas, soy foods (in moderation)
- *Beverages:* green and black tea, herbal teas, coffee
- *Dark chocolate*

Healthy Fats

As you clean out those kitchen cupboards, remove those highly processed vegetable oils like corn oil and canola oil. Yes, they're inexpensive and good for frying, but they're so heavily processed that their nutrition and flavor are mostly gone. Stick to better vegetable oil choices for dietary fat, such as olive oil, nut oils, and avocado oil. When choosing these oils, look for products that are minimally processed. Extra-virgin cold-pressed olive oil is very widely available. You may have to look in the health food aisles for other types of cold-pressed or unrefined oils, such as grapeseed oil, peanut and other nut oils, sesame oil, and sunflower oil.

I recommend staying away from coconut oil as a cooking fat. Although coconut oil is touted as being good for your gut microbiome, the evidence isn't strong—and saturated fat, even if it's from a plant food, is still bad for you. Coconut oil will raise your bad cholesterol just as any other saturated fat will. Use it only now and then for the subtle flavor it gives to South Asian dishes. If you're a vegan, coconut oil can be a good substitute for butter, but I suggest using it sparingly.

Fatty cold-water fish such as salmon, tuna, mackerel, herrings, sardines, and anchovies are rich in omega-3 fatty acids. So are caviar and oysters—woo hoo! Plant foods that are high in omega-3s include nuts (especially walnuts and pecans), flaxseeds, chia seeds, hemp seeds, pumpkin seeds, edamame (immature soybeans), tofu (made from soybeans) and other beans, especially kidney beans. Winter squashes such as acorn squash, butternut squash, pumpkin, and other hard-skinned types are also good sources.

Soothing Spices

The many different phytochemicals in herbs and spices like peppermint and turmeric not only give them their strong taste and smell but also help them counteract nausea, heartburn, and abdominal cramps. In traditional herbal medicine, anise, ginger, fennel, chamomile, lavender, lemon balm, cinnamon, turmeric, cumin, cardamom, bay leaves, and holy basil (tulsi) are all recommended for mild digestive upsets—and also as relaxing bedtime drinks. You can make any of them into a soothing hot tea or infusion. This is a fun area to explore because you can mix and match and come up with your own blends. Sweeten them with a dash of honey.

Gut Reno Fitness Plan

My approach to fitness is pretty nimble because I don't always have the time or inclination to trek to the gym. I prefer to squeeze in my workouts when and where I can. That works, because over the course of a day I can usually manage at least three, and sometimes more, short energy snacks of just ten to twenty minutes each. It all adds up to at least thirty minutes a day, almost every day, which is always the goal.

Depending on how my day is going, I split my workouts between the gym and home. If I have more time, I like to do at least thirty minutes at the gym. When I do a home workout, I incorporate the exercises in the section at the back of this book into my routine. I switch them around and try new ones just to keep things interesting and keep myself motivated. Doing the same handful of exercises all the time works, but without some fun, some variety, and new challenges, you'll find it all too easy to slack off. There are *so* many great videos to be found online.

Even if you're a gym novice, the exercises at the back of this book are good starting points for daily workouts. I know it can be hard to exercise during the workday, so I've included exercises you can subtly do at your desk without needing to put on workout clothes (though you might want to kick off your heels). The important thing is to take advantage of those brief downtime periods to work in some movement. I like to do wall presses, for instance, while I wait for the microwave to heat up my lunch. In terms of aerobic exercise, biking, spin classes, yoga, swimming, and similar exercises are great for your heart and building muscles. Make a special effort to also include weight-bearing exercise every day. These exercises, which make your body work against gravity, are vital for maintaining strong bones. Walking, jogging, running, climbing stairs, and dancing are all good examples.

Weight-training/resistance exercises are essential for strong bones and strong muscles. If you have the space, a set of dumbbells and ankle weights are great. If you don't, I recommend resistance bands instead. These are elastic bands that come in a range of sizes that correspond to dumbbell weights. They're inexpensive, lightweight, and don't take up a lot of room, so they're convenient not just at home but also at the office and when you're traveling. Another advantage is that resistance bands maintain constant tension on your muscles throughout the entire movement of an exercise, which is great for building muscles. Bonus: You can't drop a resistance band on your foot. Okay, technically you can—but it won't hurt!

As a doctor, I have to add this: if you have any sort of chronic health problem, such as back or joint pain or heart disease, discuss your exercise plans with your doctor before beginning.

STAYING HYDRATED

To make sure I start an exercise session with enough fluid in my system, I drink about 8 ounces of water half an hour before I begin. I make sure to pause for another 8 ounces or so about halfway through and finish up my workout with another drink of water. Why is hydration so important during exercise? You need water to regulate your body temperature and lubricate your joints. Your muscles are about 75 percent water—they won't work well if you don't give them the water they need. As you exercise, you lose water through sweating and by breathing hard. If you don't start out hydrated and don't replenish the lost water during your exercise session, you may have muscle cramps, get tired too soon, overheat, or feel dizzy. Your workout definitely won't be as effective or very much fun.

Unless you exercise very hard, sweat a lot, or are working out in very hot weather, plain water is fine for hydrating before and during a workout. You probably don't need a sports drink—and you also probably don't need the sweeteners and chemicals these drinks contain.

When your workout is done, have some more water. Then, within an hour of finishing, have a snack or shake containing about 20 grams of protein. Research tells us protein is particularly important for building muscle, decreasing muscle fatigue, and soothing muscle soreness. A couple of high-protein snack bars is a convenient choice. Other easy choices are nut butters, hard-boiled eggs, chocolate milk, live-culture yogurt, cottage cheese, and tuna. When I have the time, however, I like to make myself a protein smoothie with unflavored whey protein. It hydrates me, tastes great (no gritty, chemical flavor!), and gives me an energy boost—plus it's a healthy reward for exercising. My favorite basic high-protein smoothie recipe is in the recipe section at the end of this book.

Gut Reno Mind Plan

As we learned in chapter 7, "The Zen Corner," high stress is a route to disturbed digestion, both short-term and long-term. I can't promise that learning to handle stress will magically fix your digestive issues, but I can say with confidence that it will help—probably a lot. It will also help with other stress-related issues, like headaches and tight back muscles, and reduce your risk of many chronic stress-related diseases.

A daily practice of some sort of stress-relieving action helps give you a better outlook and builds the resilience you need to handle what life throws at you. But what works for stress reduction is very individual. Some people find Zen meditation helps; others prefer knitting. As with exercise, it's not what you do, it's that you do it regularly, preferably every day. Here are some of my favorite ways to combat and manage stress.

Stay mindful through apps: There are many apps designed to help you reduce stress through a variety of techniques, including mindfulness, meditation, and breathing exercises. Personal Zen, Calm, UCLA Mindful, and Simply Being are a few of my go-tos.

Try a daily gratitude check-in: Every morning when you wake up, name three things for which you are grateful. Big or small, it doesn't matter. Keeping a gratitude journal is another great way to do this, with the benefit of having something written down that you can go back to in dark moments.

Use daily affirmations: When you head to the bathroom in the morning and look in the mirror, say something kind to yourself. Rather than allowing negative self-talk to dominate your thoughts, give yourself compliments frequently.

Exercise your brain: People who stay intellectually active reduce their risk of age-related cognitive decline, so if you want to be mentally sharp when you're older, the time to start is now. Reading is an ideal brain exercise. Other forms of brain exercise include doing crossword puzzles, learning a new language, pursuing something that interests you (local history, for example), playing a musical instrument (it's never too late to learn), and pursuing a craft or hobby.

Be a social butterfly: The negative physical and mental effects of loneliness are clear, so being social is vitally important, even if you love your alone time. Social events like library book clubs or community classes like pottery or dancing are another great way to stay connected and creative!

Gut Reno Sleep Plan

Have you done a Gut Renovation of your bedroom yet? If you did, you probably now have a comfy new mattress and pillows, light-blocking window treatments, a white noise machine, and a fan or air conditioner to keep the room on the cool side. You're on your way to better sleep!

Check back to chapter 8, "The Bedroom," for all the reasons you should make good sleep a priority for your gut health and your health overall. That chapter also gives you a lot of helpful, actionable sleep hygiene tips. I strongly recommend putting them into action to help you get a solid night's sleep almost every night. Your microbiome can get seriously disrupted when your sleep is irregular, or you don't get enough of it.

SLEEP WELL ROUTINE

Stick to a schedule: My top tip for improving your digestion and your energy simultaneously is sticking to a regular schedule that gives you the quantity and quality of sleep you need to feel your best. That means going to sleep and getting up at approximately the same times every day, weekends included.

Digital detox: To help you get ready for bed, I recommend a relaxing late-evening routine that includes turning off the screens at least an hour before your bedtime. Too often, we stay up late doing things that get our mind racing, and then can't turn our thoughts off when we finally do decide to go to sleep.

Bedtime ritual: Instead of scanning your phone, read a book, take a relaxing hot bath or shower, have a hot cup of a relaxing herbal tea, meditate, or write in your journal—anything that relaxes you and calms your mind will work to send you to sleep soon after you turn out the light.

Gut Reno Beauty Regimen

In chapter 5, "The Powder Room," I laid out the intimate connection between your gut and your skin. So, if your gut is inflamed, you may see it on your skin as a breakout of acne. But on the other hand, what's good for your gut is also good for your skin. That's yet another good reason to eat plenty of those fruits and vegetables. They're rich sources of antioxidants, which help prevent wrinkles, dark spots, and other signs of aging by protecting your skin from sun damage. A healthy diet also supports your production of collagen and elastin in the skin, slowing the appearance of wrinkles and sagging.

SKIN SURVIVAL GUIDE

Here comes the sun: Protecting your skin from the damaging rays of UV from sunlight is essential. I'm a wide-brimmed-hat kind of gal, but at the very least you need to use a good sunscreen and apply it liberally—every darn day, rain or shine, winter or summer.

Feed your skin: Sunscreen isn't perfect, however, and unless you're a total night owl, you have to go outdoors in the sun sometimes.

The damage from the UV may happen before your body's natural antioxidants can prevent it. That's why those fruits and veggies are so important. Plant superfoods for healthy skin are the most colorful ones you can find: blueberries, carrots, tomatoes, spinach, sweet potatoes, and all the leafy green vegetables. Probiotic foods are great for your skin; so is wheat germ (great dietary source of vitamin E) and fish oil (natural moisturizing from the omega-3 fatty acids). At the end of this book I've included a couple of easy, skin-friendly recipes, but really, all the recipes I give are great choices for helping your skin stay in great shape.

Skin-loving ingredients: You can also apply many of these antioxidant superfoods, as well as prebiotics and probiotic extracts, directly onto your skin through serums and other products that contain them in topical form. Other key ingredients to incorporate into your daily skin care routine for hydration and defense against wrinkles are hyaluronic acid, collagen, and ceramides.

Detox gently: Another skin-damaging agent is air pollution, which is why you should look for skin care products with antipollution benefits. It's also why you should thoroughly cleanse and exfoliate your skin. Using a gentle exfoliating product once or twice a week, you can slough off the dead skin cells and debris that accumulate on your face and cause dull, tired-looking skin. You'll also absorb your serums more with a freshly exfoliated face. But the key word here is *gentle.* No cleanser or scrub should leave your skin feeling tight, dry, or irritated. Products that don't disrupt your skin's natural hydration or its microbiome allow your skin to feel soothed and supple.

Stress shows: When we suffer from stress, the hormones released can cause skin breakouts, so stick to your daily mindfulness practice to keep your skin even and blemish-free.

Beauty sleep: The best under-eye dark-circle remover is a good night's sleep—and it also improves dull skin. So make sure your bedroom routine is fully renovated, and use an extra-nourishing and hydrating night cream so that you wake up with a refreshed morning glow.

GUT RENO WORKOUT WEEK

When it comes to exercise, I like to mix it up. It keeps my workouts interesting and fun, and it also varies the muscles I use each time. Some days are heavier on the legs, for example, while others work your core more or use more upper body muscles. It all evens out over the course of the week.

I know for me it's been helpful to map out what a week of fitness looks like, so the program I outline here is just a jumping-off point. Think of it as a framework for how you can build your own workouts over the course of seven days. If you find you really hate doing a particular exercise, or if doing it is too uncomfortable or difficult for you, cross it off your list—but find something to replace it. Don't forget, there are tons of outdoor activities that count as exercise—everything from walking to gardening. It's all about choosing what works for you and ramping up from there!

If you're new to exercise, or are getting back to it after a long break, start slowly and gently. Challenge yourself, but don't overdo—gradually build up your strength, flexibility, and endurance.

Monday

30 minutes running.

12 minutes core floor exercises. Your core muscles consist of your abdominal muscles but also the obliques (the muscles on the sides of your trunk) and the muscles of your pelvis, lower back, and hips. Together, these muscles help stabilize your body. Do each of the following exercises for 45 seconds, then rest for 15 seconds, then repeat for 45 seconds.

1. *Russian twists.* Sit on your exercise mat with your knees bent and feet flat. Lean back a bit so your spine is at a 45-degree angle from the floor—you want to create a V shape with your torso and thighs. Clasp your hands together against your chest and lift your feet up a couple of inches. Using your abdominal muscles, twist to the right, then back to the center, and then to the left. Repeat. For more of a challenge, hold a dumbbell.
2. *Classic crunch.* Lie on your back on the mat with your knees bent and feet flat, both hip-width apart. Cross your arms on your chest. Inhale and contract your ab muscles. Lift your upper body, keeping your head and neck relaxed (don't lift with them) while exhaling. Inhale as you lower yourself back to the starting position.
3. *Bicycle crunch.* Lie on your back on the mat with your knees bent and feet flat, hip-width apart. Place your arms behind your head, with your elbows pointing out. To start, inhale and brace your abs. Lift your knees to 90 degrees and raise your upper body up. Exhale and rotate your trunk, moving your right elbow and left knee toward each other while also straightening your right leg. Hold for 5 seconds, then inhale as you return to the starting position. Repeat on the other side: move your left elbow to your right knee and extend your left leg. Pause and return to the starting position. Note: keep your lower back on the floor and don't hunch your shoulders. Rotate from your core, not your neck.
4. *Reverse crunch.* Lie on your back on the mat with your knees bent and feet flat, hip-width apart. Put your arms at your sides, palms down.

Exhale, brace your core, and slowly lift your feet off the floor, raising your thighs until your legs are vertical (or as near vertical as you can get). Pull your knees in toward your head as far as you can without lifting your middle back from the mat—your hips and lower back should be off the mat. Hold for 5 seconds, then slowly lower your feet down to the starting position.

5. *Straight-arm plank.* Get on all fours with your wrists aligned under your shoulders. Step one foot back and then the other until your legs are straight with your toes on the mat. Press down on your hands to raise your body into a straight line from your shoulders to your heels, as if you were in the up position of a push-up. Look down and keep your spine straight—don't hunch your back or your shoulders. Hold for as long as you can with good form, then lower yourself back down.

6. *Mountain climbers.* Start in the straight-arm plank position. Pull one knee into your chest as far as you can, then extend it again while exhaling. Switch legs and repeat. Once you get the hang of this, do it as fast as you can. Be sure to keep your hips and butt down and your shoulders over your wrists.

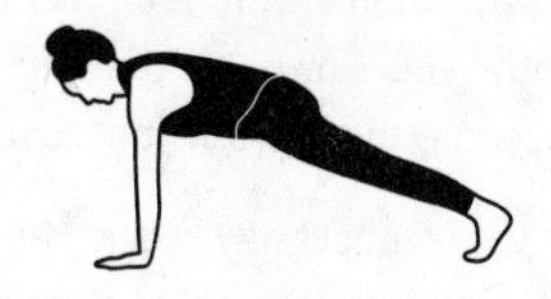

Tuesday

30 minutes aerobic dance video (belly dance, hip hop, and Bollywood are some of my faves).

12 minutes weight-bearing (bone strength) exercises. For these exercises, you'll need a straight-backed chair for support. Do 10 reps (1 set) of each exercise, resting for 1 minute between sets, then repeat routine. Note: these exercises are particularly good for short exercise snacks at the office.

1. *Sit to stand.* Stand with the chair behind you, bend at the knees and hips, and slowly lower yourself to a seated position, then slowly stand up.
2. *Forward lunge.* Stand with your feet shoulder-width apart. Step forward with one foot, planting it firmly on the ground. Slowly shift your weight to your front foot, lowering your body as you do so. Return to starting position.
3. *Single leg standing.* Stand with your feet shoulder-width apart. Bend one knee slightly, and slowly lift that leg 3 to 6 inches off the floor. Hold for 10 seconds, then return your foot to the floor.
4. *Squats.* Stand with your feet slightly wider than shoulder-width apart. Bend downward at the hips, shifting them back and down as you bend your knees. Lower yourself to a comfortable position—your knees shouldn't extend beyond your toes. Push through your heels to return to the starting position.

Wednesday

30 minutes swimming or running.

10 minutes yoga stretches. Hold each pose for 45 seconds with 15 seconds rest in between. Repeat routine 1 time. Yoga poses are best learned by having someone teach them to you. I recommend taking a few basic yoga classes or watching one of the many excellent yoga videos online. There are hundreds of yoga poses. The ones below are good basic starting points for developing your own practice.

1. *Mountain pose*
2. *Standing forward bend*
3. *Downward dog*
4. *Tree pose*
5. *Upward salute*
6. *Triangle pose*
7. *Warrior pose*
8. *Side-angle pose*
9. *Twisted triangle*
10. *Locust pose*

Thursday

30-minute HIIT video (Popsugar has a great one).

10 minutes balance exercises. Do each exercise for 1 minute, rest for 15 seconds, then repeat. Note: these exercises are particularly good for short exercise snacks at the office.

1. *In-place marches.* Stand upright with feet hip-width apart. Lift one knee until your thigh is parallel with the floor. Pause, then slowly return your foot to the floor. Alternate your right and left legs.
2. *Tightrope walk.* Stand up straight with your arms straight out from your sides. Walk heel to toe in a straight line—one foot touching right in front of the other, pausing for 2 seconds every time you lift one foot off the ground.
3. *Quad stretch.* Stand upright with your feet hip-width apart. Balancing on your right leg, grab your left ankle from behind with your left hand and pull your foot up to meet your butt. Hold for one minute, then switch to the other leg. If you can't lift your leg that high, just bring it up as far as you can without losing your balance.
4. *Head rotations.* Stand upright with your feet hip-width apart. Slowly move your head from side to side, then up and down for 30 seconds. Move your head around clockwise for 15 seconds and then counterclockwise for 15 seconds.

Friday

30 minutes biking (real or stationary).

10 minutes core exercises. Do each exercise for 45 seconds, rest for 15 seconds, then repeat. Once you get to the point where doing these isn't that hard, don't do them for longer periods of time. Instead, use resistance bands to make them more challenging.

1. *Heel taps.* Lie on your back with your hands under your butt, knees bent, and feet lifted into a tabletop position. Flex your feet and slowly lower them to the ground until your heels barely touch the floor. Squeezing your abs, lift your feet back up to starting tabletop position.
2. *Scissor kicks.* Lying on your back, lift your head and shoulders off the floor. Lift your right leg until it is at about a 45-degree angle from your body and then lower it. Switch legs and repeat. Keep switching for 45 seconds.
3. *Straight leg raises.* Lie on your back. Breathing in and tightening your abs, raise both legs (keeping them straight) until they're perpendicular to your torso (or as perpendicular as you can get them). Then exhale and slowly lower your legs until they're a few inches above the floor (or as close to that as you can get without lifting your lower back from the floor).
4. *Roll ups.* Lie on your back with your arms and legs outstretched. As you inhale, bring your arms overhead and slowly start to curl your upper body off the floor. Keep rolling forward to reach your toes (or as close to your toes as you can get). Then reverse the move as you exhale, allowing one vertebra at a time to rest back down on the ground.
5. *High knees.* Stand with your feet hip distance apart, then start to run in place, lifting your knees up in front of you as high as they'll go. As you pump your legs, swing your opposite arm to give yourself more momentum.

Saturday

30 minutes brisk walking.

10 minutes arm exercises. Do 15 reps (1 set) of each exercise using 5-pound hand weights, dumbbells, or resistance bands. Rest 30 seconds between sets and repeat. As you build strength, use heavier weights—don't do more reps. Breathe out as you lift the weights and in as you lower them.

1. *Biceps curls.* Hold the dumbbells parallel to the floor and stand up straight, your legs shoulder-width apart and your elbows pressed against your sides. Raise the dumbbells to the count of 3 by bringing your forearms up to the front of your shoulders, rotating the dumbbells so your palms face your shoulders. Pause for a breath, then lower the dumbbells to the count of 3 to the starting position.
2. *Overhead press.* Stand up straight with your legs shoulder-width apart and the dumbbells held up at shoulder height with your palms facing forward. Raise the dumbbells to the count of 3 by bringing your arms straight up over your head. Pause for a breath, then lower the dumbbells to the count of 3 to the starting position.
3. *Upright row.* Stand up straight with your legs shoulder-width apart and the dumbbells resting on the fronts of your thighs, with your palms facing inward. Pull the dumbbells up to the count of 3 until they're just below your chin. Pause for a breath, then lower the dumbbells to the count of 3 to the starting position.
4. *Triceps extension.* Stand up straight with your legs shoulder-width apart and hold a single dumbbell in the center with both hands. Extend your arms up over your head with both hands, keeping your arms close to your ears and your elbows pointed forward. Lower the weight to the count of 3 behind your head until your elbows are at about 90-degree angles. Pause for a breath, then raise your arms up again to the count of 3 to the starting position.

5. *Side (lateral) raises.* Stand up straight with your legs shoulder-width apart and hold a dumbbell in each hand with your arms at your sides, palms facing in. Raise both arms out and up to the sides to the count of 3 until your elbows are shoulder-height. Pause for a breath, then lower your arms to the count of 3 to the starting position.

Sunday

Take a day off—you've earned it!

GUT RENO RECIPES

I wear a lot of hats: mom, doctor, entrepreneur. Ergo, I don't have a lot of time to spend in the kitchen. And while I have a pretty comprehensive résumé, a stint at the Cordon Bleu is conspicuously absent. All the recipes here are ones that meet my basic criteria for entering the permanent repertoire: they taste great, they're good for your gut, they're quick and easy to make, they don't require a lot of expensive or hard-to-find ingredients, and my boys eat them without complaint (usually).

Breakfast

Breakfast is a good time to get started on your daily fiber and prebiotics servings. For quick, no-fuss breakfasts, try oatmeal or a bowl of commercial whole-grain or bran cereal with milk or yogurt, topped with berries or other fruit. Look for cereal brands that contain 100 percent whole grains, have at least 5 grams of fiber per serving, and don't have added sugar. For more variety, try low-sugar granola. Check the label: the first ingredient listed should be whole grains.

Probiotic Parfait

This is my usual workday quick breakfast. The key ingredients are the yogurt and granola; beyond that, anything prebiotic goes. I like to toss in some sliced toasted almonds.

Makes 1 serving

INGREDIENTS

- ½ cup low-sugar granola
- 1 cup plain nonfat Greek yogurt
- 1 tablespoon flaxseeds
- 1 tablespoon chia seeds
- ½ cup berries (any kind or a mixture)

PREPARATION

Put the granola in a cereal bowl, add the yogurt, top with the seeds and berries.

Avocado Toast

This basic recipe becomes a heartier breakfast or a lunch dish when you top it with sliced veggies, a fried or hard-boiled egg, salsa, or some feta cheese.

Makes 1 serving

INGREDIENTS

- 1 ripe avocado
- ¼ teaspoon lemon juice
- ¼ teaspoon salt
- ¼ teaspoon black pepper
- 2 slices whole-grain bread
- 1 teaspoon extra-virgin olive oil
- pinch of cumin

PREPARATION

1. Scoop out the avocado into a small bowl. Add the lemon juice, salt, and black pepper and mash coarsely with a fork.
2. Toast the bread to your preferred darkness. Spread the avocado mix on the bread, drizzle with the olive oil, and sprinkle with cumin.

Overnight Oats

The quickest, easiest, healthiest breakfast I know. Top the oatmeal with anything you like, singly or in combination. Berries, bananas, apple or peach slices, raisins, nuts, and shredded coconut all work. To add some sweetness, try some maple syrup or honey. My personal favorite topping is a spoonful of almond butter, some banana slices, and a sprinkle of cocoa nibs. You can multiply the recipe by as many servings as you need for family breakfast but use them the next morning or the day after for best flavor and texture.

Makes 1 serving

INGREDIENTS

½ cup dairy milk, nut milk, or oat milk

½ cup old-fashioned rolled oats (don't use quick or instant oats)

½ cup plain nonfat Greek yogurt

1 teaspoon chia seeds

PREPARATION

1. Combine all ingredients in a Mason jar or other airtight container, stir well, cover, and refrigerate for at least 5 hours, preferably overnight.

2. To serve, stir in some extra milk or yogurt if desired and top with anything you want.

Buckwheat Pancakes

The rich, nutty flavor of these fiber-filled buckwheat pancakes will make you rethink your commitment to regular pancakes. And if you love pancakes but can't tolerate the gluten, buckwheat is the grain for you. Important note: your pancake griddle needs to be really hot before you start cooking the pancakes. This recipe is so quick and easy that I recommend heating the griddle while you prepare the batter.

Makes 6 pancakes

INGREDIENTS

- ½ cup buckwheat flour
- ½ teaspoon baking powder
- ⅛ teaspoon salt
- ½ cup milk or unsweetened milk substitute
- 3 tablespoons maple syrup

PREPARATION

1. Combine the buckwheat flour, baking powder, and salt in a medium mixing bowl. Whisk in the milk and maple syrup.
2. If using a regular griddle, lightly grease it with a small amount of neutral oil (grapeseed works well). Drop the batter by large spoonfuls onto the griddle and cook until the entire surface is covered with dry bubbles, about 2 to 4 minutes. Flip and cook for an additional 2 to 4 minutes.

Healthy Lunches

These quick and easy lunches are great on weekends and make a good alternative to fast food or deli meats.

Vegetable Frittata

A frittata is best served warm or at room temperature, making it ideal for make-ahead lunches. This recipe gives you the general idea, but use any mixture of veggies you like.

Makes 4 servings

INGREDIENTS

- 10 eggs
- ½ cup dairy milk
- ½ teaspoon salt
- ¼ teaspoon black pepper
- 2 tablespoons extra-virgin olive oil
- ½ red onion, thinly sliced
- 4 ounces white or baby brown mushrooms, thinly sliced
- 1 medium red bell pepper, seeded and thinly sliced
- 4 asparagus spears, cut into ½-inch pieces
- 4 ounces feta cheese

PREPARATION

1. Preheat the oven to 350°F.
2. Whisk together the eggs, milk, salt, and pepper.
3. Heat the olive oil in a large, oven-proof skillet. Add the red onion and mushrooms and cook, stirring now and then, for 4 minutes or until the mushrooms start to take on color. Add the pepper and asparagus spears and cook for another 4 minutes.
4. Pour the egg mixture over the veggies and cook without stirring for another few minutes, or until the eggs begin to set around the edges.
5. Sprinkle the feta cheese over the eggs. Put the skillet into the oven and bake for 20 minutes. Remove from the oven and let rest for 5 minutes before serving.

Healthy Tuna Salad

This quick sandwich lunch substitutes yogurt for mayo. The microgreens add additional fiber and flavor—use whatever greens are in season. Serve on whole-grain bread or in a whole-wheat wrap.

Makes 4 servings

INGREDIENTS

- 2 5-ounce cans water-packed tuna, drained
- 1 celery stalk, diced
- 2 tablespoons red onion, diced
- 1 cup microgreens, such as baby arugula or radish sprouts

Dressing

- ⅓ cup plain nonfat Greek yogurt
- 2 tablespoons lemon juice
- 1 tablespoon Dijon or yellow mustard
- ¼ teaspoon salt
- ¼ teaspoon black pepper
- 2 tablespoons chopped parsley

PREPARATION

In a medium bowl, combine the yogurt, lemon juice, mustard, salt, pepper, and parsley. Add the tuna, the celery, red onions, and microgreens. Stir gently to combine.

Portobello Sandwich

Meaty portobello mushrooms are great in sandwiches, a much better alternative to lunch meats. Make this sandwich on whole-grain bread or rolls brushed with olive oil. Add a side of sauerkraut or pickles for a dose of probiotics.

Makes 4 servings

INGREDIENTS

4 large portobello mushroom caps

1 tablespoon extra-virgin olive oil

1 teaspoon salt

1 teaspoon dried thyme

½ teaspoon garlic powder

1 large tomato, thickly sliced

4 slices avocado

PREPARATION

1. Brush the mushroom caps with half the olive oil and sprinkle with salt, thyme, and garlic powder.
2. Add the remaining olive oil to a skillet and heat over medium heat. Place the mushrooms in the skillet, round side down, and cook for 5 minutes.
3. Make sandwiches with the caps and top with tomato slices and avocado slices.

Salads

Basic Vinaigrette Dressing

If you want more fiber in your diet, eat more salad. Make your salads more interesting by using a variety of different lettuces and baby greens and adding whatever other raw vegetables you like. Add some additional crunch with sunflower seeds or chopped nuts. Because bottled salad dressings often contain added sugar and other unhealthy ingredients, learn to make your own, starting with this basic vinaigrette.

Makes 1 cup

INGREDIENTS

- 3 tablespoons balsamic, red wine, or cider vinegar
- 1 garlic clove, finely chopped
- 1 teaspoon Dijon or whole-grain mustard
- 1 teaspoon kosher salt
- ½ teaspoon black pepper
- ¾ cup extra-virgin olive oil

PREPARATION

Combine all the ingredients in a pint jar with a lid. Shake well to combine. Store the dressing in the refrigerator. Let the dressing come to room temperature before serving.

Watermelon, Spinach, and Tomato Salad

This is the ultimate summer salad. The watermelon, spinach, and tomatoes are all excellent sources of lycopene, a natural antioxidant that helps protect your skin from sun damage.

Makes 4 servings

INGREDIENTS

Dressing

3 tablespoons extra-virgin olive oil

1 tablespoon apple cider vinegar

½ teaspoon kosher salt

Salad

4 cups fresh spinach leaves

1 cup thinly sliced red onion

1 cup cherry tomatoes, cut in half

2 cups watermelon chunks

PREPARATION

1. Whisk the olive oil, apple cider vinegar, and salt together in a small bowl.
2. Combine the spinach, red onion, and tomatoes in a salad bowl. Add the vinaigrette and toss to coat. Add the watermelon just before serving and toss gently again.

Quinoa Salad with Artichokes, White Beans, and Pistachios

The pistachios add a nice crunch to this salad. Quinoa is a great source of fiber—it has twice as much as most grains.

Makes 4 servings

INGREDIENTS

1 cup quinoa

1 15-ounce can small white beans or navy beans

½ cup marinated artichoke hearts, chopped

1 cup cherry tomatoes, cut in half

¼ cup red onion, chopped

½ cup pistachios

Dressing

3 tablespoons extra-virgin olive oil

1 tablespoon lime juice

1 clove garlic, minced

¼ teaspoon cumin

¼ teaspoon cayenne pepper

1 teaspoon salt

PREPARATION

1. Combine the quinoa with 2 cups water in a medium saucepan over medium heat. Bring to a boil, reduce the heat to a simmer, cover the pot, and cook for 15 to 20 minutes, or until all the water is absorbed.

2. While the quinoa cooks, drain the beans and rinse them in a colander. Prepare the artichokes, cherry tomatoes, and red onions.

3. For the dressing, whisk the olive oil, lime juice, garlic, cumin, cayenne pepper, and salt together in a small bowl.

4. When the quinoa is ready, fluff it with a fork and put it into a large serving bowl. Gently mix in the beans, artichokes, cherry tomatoes, red onions, and pistachios. Drizzle with the dressing and let sit for 5 minutes before serving so the flavors can blend.

Corn and Avocado Salad

Makes 4 servings

INGREDIENTS

1½ cups corn kernels

1 avocado, cut into small cubes

½ cup thinly sliced red onion

2 jalapeño peppers, seeded and finely chopped

½ cup coarsely chopped cilantro leaves

Dressing

3 tablespoons extra-virgin olive oil

1 tablespoon balsamic vinegar

½ teaspoon cumin

½ teaspoon salt

PREPARATION

1. Combine the corn kernels, avocado cubes, red onion, jalapeños, and cilantro in a serving bowl.
2. Whisk together the olive oil, vinegar, cumin, and salt in a small bowl. Pour the dressing over the corn and avocado mixture and toss very gently. Let stand at room temperature for 30 minutes before serving so the flavors can blend.

Vegetable Side Dishes

Lemony Broccoli Rabe with White Beans

Broccoli rabe (also called rapini) has a bitter flavor when raw but mellows out beautifully when cooked.

Makes 4 servings

INGREDIENTS

- 1 pound broccoli rabe
- 2 tablespoons extra-virgin olive oil
- 1 lemon, very thinly sliced
- 1 large garlic clove, chopped
- 1 15-ounce can cannellini or navy beans, drained and rinsed
- ½ teaspoon hot red pepper flakes
- ½ teaspoon salt
- 1 tablespoon grated Parmesan

PREPARATION

1. Trim the broccoli rabe and cut it into 4-inch pieces. Cut the thicker stalks in half lengthwise so all the broccoli rabe cooks evenly.
2. Heat the olive oil in a large skillet over medium heat. Add the lemon slices in an even layer and cook for 2 minutes. Turn the slices and cook for another 2 minutes. Add the broccoli rabe pieces and garlic and cook, stirring occasionally, about 5 minutes, or until the pieces are bright green and tender.
3. Add the beans, red pepper flakes, salt, and ½ cup water. Bring to a boil, lower the heat, and simmer for another 5 minutes or so, stirring occasionally, until the liquid is reduced by half. Stir in the Parmesan.

Roasted Turmeric Chickpeas

One of my favorite side dishes and a great source of fiber and probiotics.

Makes 4 servings

INGREDIENTS

- 2 15-ounce cans chickpeas
- 3 tablespoons extra-virgin olive oil
- 1 teaspoon turmeric
- 1 teaspoon fennel seeds
- ½ teaspoon salt
- ½ teaspoon ground black pepper
- ½ cup plain nonfat Greek yogurt
- 4 tablespoons lemon juice
- 1 teaspoon hot red pepper flakes (optional)

PREPARATION

1. Preheat the oven to 400°F. Drain the chickpeas in a colander, rinse, and spread them on towels to dry.
2. Combine the olive oil, turmeric, fennel seeds, salt, and black pepper in a medium mixing bowl.
3. Add the chickpeas and toss well to coat.
4. Spread the chickpeas in a single layer on a nonstick baking tray or sheet pan. Bake for 20 to 30 minutes, or until the chickpeas are golden brown and a bit crispy.
5. Remove from the oven and let cool. Put the chickpeas in a medium mixing bowl and stir in the yogurt, lemon juice, and red pepper flakes if desired.

Sweet Potatoes and Peppers

Makes 4 servings

INGREDIENTS

2 large sweet potatoes

1 large red bell pepper

1 large green bell pepper

1 medium red onion

2 tablespoons extra-virgin olive oil

2 teaspoons dried thyme

2 teaspoons smoked paprika

1 teaspoon hot red pepper flakes

1 teaspoon salt

PREPARATION

1. Preheat the oven to 425°F.
2. Peel the sweet potatoes and cut them into 1-inch chunks. Seed the peppers and cut them into thin slices. Cut the red onion into 1-inch chunks.
3. Put the olive oil into a medium mixing bowl. Add the thyme, paprika, red pepper flakes, and salt and stir well. Add the sweet potatoes, red and green peppers, and the onion. Toss to coat them with the spice mixture.
4. Spread the vegetables out on a sheet pan and roast for 15 minutes. Use a spatula to turn the vegetables and roast for 15 minutes longer, or until the sweet potatoes are lightly browned and soft.

Carrots with Cumin

Makes 4 servings

INGREDIENTS

1 pound carrots

1 teaspoon cumin seeds

1 large garlic clove, finely chopped

¼ cup extra-virgin olive oil

1 cup orange juice

½ teaspoon salt

1 teaspoon sugar

2 tablespoons chopped cilantro

1 teaspoon lemon juice

PREPARATION

1. Peel and trim the carrots. Cut them into ½-inch slices.
2. Combine the cumin seeds, garlic, olive oil, orange juice, salt, and sugar in a medium saucepan. Bring the mixture to a boil over medium heat, stirring as you do.
3. Add the carrots, cover the saucepan, and reduce the heat to very low. Continue to cook, stirring now and then, for 30 minutes, or until the carrots are very soft. Stir in the cilantro and lemon juice and serve.

Sweet Potato Fries

As simple, tasty, and healthy as it gets.

Makes 4 servings

INGREDIENTS

- 4 large sweet potatoes
- 2 tablespoons extra-virgin olive oil
- 1 teaspoon garlic powder
- 1 teaspoon paprika
- 1 teaspoon sea salt
- ½ teaspoon black pepper

PREPARATION

1. Preheat the oven to 400°F.
2. Peel the sweet potatoes and cut them into sticks about ½ inch wide and 3 inches long. Don't discard any smaller leftover pieces—just cook them along with the sticks.
3. Combine the olive oil with the garlic powder, paprika, salt, and black pepper in a medium mixing bowl. Add the sweet potato sticks and any smaller pieces and toss to coat them with the olive oil mixture.
4. Spread the pieces evenly on a nonstick baking sheet (you might need two).
5. Bake until browned and crisp on the bottom, about 15 minutes. Flip the sticks over and bake until the other side is browned, about 10 minutes more.

Baked Zucchini Fries

A great alternative to low-fiber french fries. You can substitute yellow summer squash for the zucchini or use a mixture of the two.

Makes 4 servings

INGREDIENTS

- 4 medium zucchini
- 1 cup panko breadcrumbs
- ½ cup grated Parmesan
- 1 teaspoon garlic powder
- 1 teaspoon dried oregano
- 1 teaspoon dried parsley
- ½ teaspoon dried basil
- ½ teaspoon hot red pepper flakes
- ½ teaspoon salt
- ½ teaspoon black pepper
- 2 eggs

PREPARATION

1. Preheat the oven to 425°F.
2. Trim the zucchini and cut them lengthwise into quarters. Cut each quarter in half.
3. Combine the breadcrumbs, Parmesan, garlic powder, oregano, parsley, basil, red pepper flakes, salt, and black pepper together in a shallow dish.
4. Whisk the eggs in a shallow bowl.
5. Dip the zucchini sticks into the egg wash, then roll them firmly in the panko mixture to coat and lay them on a nonstick baking sheet.
6. Bake for 15 minutes, then flip and bake 10 minutes longer, until the sticks are golden-brown and crispy.

Simple Kimchi

A staple of Korean cooking, kimchi is spicy fermented cabbage. It's a wonderful source of both probiotics from the bacteria and prebiotics from the fiber in the cabbage and other vegetables. Make a big batch—it keeps in the refrigerator for three to four months. The hotness of kimchi comes from liberal amounts of gochugaru—Korean red pepper flakes. If you can't find these, substitute Aleppo red pepper flakes. Do not use the standard hot red pepper flakes from the supermarket. You'll need a 1-quart jar with a lid to ferment the kimchi. A Mason jar is great, but really any glass container with a screw-top lid will work.

Makes about 1 quart

INGREDIENTS

- 1 medium head Napa cabbage, about 2 pounds
- ¼ cup kosher salt
- 6 large garlic cloves, finely chopped
- 2 teaspoons grated ginger
- 1 to 5 tablespoons Korean red pepper flakes
- 1 daikon, peeled, halved lengthwise, and thinly sliced
- 4 scallions, trimmed and cut into 1-inch pieces

PREPARATION

1. Cut the cabbage lengthwise into quarters and remove the core. Cut the cabbage crosswise into strips about 2 inches wide.
2. Place the cabbage pieces in a large mixing bowl and sprinkle with the salt. Use your hands to work the salt into the cabbage. Add enough cold water to just cover the cabbage. Put a plate on top and weigh it down with something heavy, like a large can of tomatoes or a pitcher of water. Let stand for 1 to 2 hours. The cabbage will give off a lot of liquid.
3. Drain the salted cabbage in a colander, pressing down to get out as much liquid as possible. While it drains, prepare the hot pepper mixture. In the large mixing bowl you used to salt the cabbage, combine the garlic, ginger, and 2 tablespoons water and stir into a paste. Add the hot pep-

per flakes to taste and stir again. Add the drained cabbage, daikon, and scallions and mix thoroughly to coat the vegetables with the spice paste.

4. Pack the kimchi into the quart jar, pressing down to remove air bubbles. Leave an inch of headspace in the jar. Cover the jar loosely and put it on a plate to catch any overflow from the brine as the kimchi ferments. Seal the jar and put it in a cool, dark place. Check the jar daily. As it ferments, you'll see tiny bubbles forming in the brine; some may overflow the jar and you'll note a definite pungent aroma. If necessary, push down on the vegetables to keep them submerged in the brine. You may see a thin white film on top. This is a yeast known as kahm and is a normal part of the fermentation process. Just skim it off and make sure your veggies are submerged.

5. Taste the kimchi after a few days. When it's fermented enough for your taste, put the jar in the refrigerator. You can eat the kimchi right away, but it's better if you let it sit for at least another week. It will keep for at least three months.

Kimchi Cauliflower Rice

Cauliflower rice has become so popular that you can now find frozen bags of it in any supermarket. It makes a quick substitute for traditional rice and is an excellent source of prebiotic fiber. If you want to make your own, use a box grater or food processor to grate a head of cauliflower into rice-like granules. Cauliflower rice cooks very quickly. Don't defrost the frozen kind before using. When using fresh or frozen cauliflower rice, cook until it's just heated through. In this basic recipe, you can substitute or add vegetables, such as grated carrots, broccoli florets, or edamame. If you don't have or don't like cauliflower rice, use 3 cups of cold brown rice instead.

Makes 4 servings

INGREDIENTS

1 tablespoon cold-pressed peanut oil or grapeseed oil

2 large garlic cloves, finely chopped

1 cup kimchi

½ cup frozen peas

2 tablespoons soy sauce

3 cups cauliflower rice or brown rice

3 eggs, beaten

PREPARATION

1. Heat the oil in a large skillet over medium heat. Add the garlic and sauté for a minute or two, then add the kimchi, peas and any other vegetables, and the soy sauce. Cook, stirring occasionally, for 3 minutes.

2. Add the cauliflower rice and cook for only 1 minute. Push the vegetable mixture to the sides of the skillet and pour the beaten eggs into the hole in the center. Cook for 1 to 2 minutes, then use a spatula to stir the eggs and mix them into the vegetables.

Fish and Seafood

For the sake of the planet and your own health, choose fish and seafood carefully. Whenever possible, look for labeling that indicates the fish meets Marine Stewardship Council (MSC) sustainability standards for wild-caught fish or has a green label from the Monterey Bay Aquarium Seafood Watch Program. For farmed fish and shrimp, look for labeling indicating it was responsibly farmed. Don't be fooled by farmed fish and seafood labeled "organic." So far, the USDA hasn't established organic standards for aquaculture.

Miso-Glazed Salmon

All cold-water fish, salmon included, are great for omega-3 fatty acids. Serve this with stir-fried baby bok choy or snap peas and brown rice for extra fiber. Miso (a great source of probiotics) is a salty paste made from fermented soybeans. It's what gives a lot of Japanese cooking its deep umami flavor.

Makes 4 servings

INGREDIENTS

4 6-ounce salmon fillets, skin on
Salt
Black pepper
4 teaspoons maple syrup or honey
2 tablespoons white or yellow miso
1 tablespoon rice wine vinegar
2 teaspoons dark soy sauce
1 large garlic clove, finely chopped

PREPARATION

1. Heat the oven to 400°F. Line a sheet pan with aluminum foil or use a nonstick pan.
2. Season the salmon fillets with salt and black pepper and place them in a shallow bowl or baking dish.

3. In a small bowl, whisk together the maple syrup or honey, miso, vinegar, soy sauce, and garlic. Pour the marinade over the fish fillets. Let marinate for 10 minutes.

4. Place the salmon fillets skin-side down on the sheet pan and bake until the salmon is opaque and flakey, about 12 minutes.

Fish Kebabs with Fennel

This recipe works well with any firm fish, such as salmon, Pacific cod, or Arctic char. Fennel is a great source of fiber. Because I like to keep things really simple in the kitchen, this recipe calls for baking the kebabs, but you could put them on the grill or under the broiler instead. And if you don't have kebab skewers, just combine all the ingredients in one layer and bake them that way.

Makes 4 servings

INGREDIENTS

- 4 6-ounce fish fillets
- 1 large fennel bulb
- 2 lemons
- 1 medium red onion
- 2 large garlic cloves, finely chopped
- 3 teaspoons hot red pepper flakes
- 4 tablespoons olive oil
- 1 teaspoon salt
- Black pepper

PREPARATION

1. Cut the fish into bite-sized chunks and place them in a large mixing bowl.
2. Trim the fennel bulb, remove the hard inner core, and cut the bulb into bite-sized pieces. Add them to the bowl with the fish.
3. Cut the lemons into thin slices. Cut the red onion into quarters and separate the layers. Add the lemons and red onion to the mixing bowl. Add the garlic, red pepper flakes, olive oil, salt, and a generous amount of black pepper. Gently mix the ingredients to coat the fish and fennel chunks.
4. Preheat the oven to 325°F.
5. Assemble the kebabs: using metal skewers, add alternating chunks of fish, fennel, lemon, and red onion until all the ingredients are used. Put the kebabs onto a sheet pan or baking dish and bake for 6 minutes. If the fish isn't flaky and done by then, cook 2 minutes longer.

Fish Tacos with Mango Salsa

Use a flaky fish such as tilapia, Pacific cod, or pollock for these simple tacos.

Makes 4 servings

INGREDIENTS

1 teaspoon cumin

1 teaspoon smoked paprika

½ teaspoon ancho chili powder

1 teaspoon salt

4 6-ounce fish fillets

1 tablespoon lime juice

2 tablespoons extra-virgin olive oil

Mango salsa

1 large mango, diced

2 tablespoons diced red onion

1 tablespoon chopped cilantro

1 jalapeño pepper, seeded and diced

1 tablespoon lime juice

½ teaspoon salt

Taco toppings

12 corn tortillas

1 cup thinly sliced red or white cabbage

2 avocados, thinly sliced

PREPARATION

1. Preheat the oven to 375°F.
2. Combine the cumin, smoked paprika, ancho chili powder, and salt in a small bowl. Mix well and rub the mixture into the fish. Sprinkle with the lime juice and drizzle with the olive oil.
3. Put the fish fillets into a baking dish and bake for 10 minutes, or until the fish is flaky.
4. While the fish cooks, make the mango salsa. Combine the mango, red onion, cilantro, jalapeño, lime juice, and salt and toss gently.
5. When the fish and salsa are ready, assemble the tacos. Place a third of a fish fillet on a corn tortilla and top with cabbage, avocado slices, and mango salsa.

Main Dishes

Stir-Fried Tofu with Cauliflower

Celery and cauliflower give this dish a hefty dose of fiber.

Makes 4 servings

INGREDIENTS

- 12 ounces extra-firm tofu
- 3 tablespoons cornstarch
- 1 cup chicken broth or vegetable stock
- 3 tablespoons soy sauce
- 1 tablespoon rice wine vinegar
- 2 teaspoons hoisin sauce
- ½ teaspoon hot red pepper flakes
- 1½ tablespoons ketchup
- 2 tablespoons cold-pressed peanut oil
- 3 cups cauliflower florets
- 2 celery stalks, thinly sliced on the diagonal
- 6 garlic cloves, thinly sliced
- ½ cup thinly sliced scallions

PREPARATION

1. Drain the tofu and cut it into 1-inch cubes. Drain on paper towels. Place 1½ tablespoons of the cornstarch in a shallow dish and toss the tofu cubes to coat.
2. In a small mixing bowl, combine 1½ tablespoons cornstarch with ¼ cup chicken broth and whisk until smooth. Add the rest of the chicken broth, soy sauce, vinegar, hoisin sauce, red pepper flakes, and ketchup and whisk until smooth.
3. Heat the peanut oil in a large skillet over high heat. Add the tofu cubes and cook until they are crispy and golden-brown, about 6 minutes. Put the cubes aside on a plate.
4. Add the cauliflower to the skillet and cook, stirring often, for 3 minutes, or until the florets are lightly browned in spots. Add the celery and garlic and cook, stirring often, for another 2 minutes. Add the cornstarch mixture

and continue to cook, stirring often, until the mixture starts to thicken. Gently stir in the tofu cubes and continue to cook for 1 minute more or until the tofu is heated through. Sprinkle in the scallions.

Eggplant Pizza

This recipe works best with large eggplants. The skin on large eggplants can be tough, however, so I recommend trimming it off.

Makes 4 servings

INGREDIENTS

- 1 large eggplant, about 10 inches long
- Kosher salt
- 2 tablespoons extra-virgin olive oil
- 1½ cups marinara sauce
- ½ cup basil leaves, coarsely chopped
- 1 teaspoon hot red pepper flakes
- 4 ounces grated Parmesan
- 4 ounces shredded mozzarella

PREPARATION

1. Cut the eggplant crosswise into circles about ½ inch thick. Place the slices on a double layer of paper towels and sprinkle with kosher salt to release the extra liquid. Let sit for 30 minutes. Rinse the slices well and pat dry.
2. Preheat the oven to 375°F.
3. Brush the eggplant slices on both sides with olive oil and arrange them on a nonstick baking sheet. Roast the slices for 15 to 20 minutes.
4. Remove the eggplant from the oven and preheat the broiler to high.
5. Spread a few tablespoons of the marinara sauce on top of each eggplant slice. Sprinkle with basil and red pepper flakes and then top each slice with the Parmesan and mozzarella.
6. Put the slices under the broiler for 5 to 10 minutes, or until the mozzarella cheese is melted and slightly browned.

Turkey Chili

A family favorite. Modify the seasoning to make it hotter by adding more jalapeños. If you like your chili soupier, add some water or chicken broth along with the diced tomatoes. I like to make a double batch and freeze half for a quick dinner later on.

Makes 4 servings

INGREDIENTS

1 tablespoon extra-virgin olive oil

1 pound ground turkey

1 cup diced onions

1 tablespoon chopped garlic

1 large red bell pepper, seeded and coarsely chopped

1 jalapeño pepper, seeded and coarsely chopped

2 teaspoons dried oregano

1 bay leaf

1 tablespoon chili powder

1 teaspoon cumin

1 15-ounce can diced tomatoes

1 teaspoon salt

½ teaspoon black pepper

1 15-ounce can red kidney beans or pinto beans, drained and rinsed

PREPARATION

1. Heat the olive oil over high heat in a large saucepan. Add the turkey and cook, stirring often to break up the lumps, for 5 minutes, or until the meat is lightly browned.

2. Add the onions, garlic, red pepper, jalapeño pepper, oregano, bay leaf, chili powder, and cumin. Stir well and cook for 5 minutes.

3. Add the tomatoes, salt, and black pepper. Bring the mixture to a boil, then reduce the heat to low and simmer, stirring occasionally, for 15 minutes. Add a small amount of water if the mixture starts to get too dry or if you like your chili soupier.

4. Add the beans and cook, stirring occasionally, for 10 more minutes.

Pasta with Mushrooms and Swiss Chard

Yes, you can make a great pasta dish with just one pot. Swiss chard is an excellent source of fiber and gives you tons of vitamin A, vitamin K, and minerals such as iron and potassium. Use baby bella (also called cremini) mushrooms for at least 4 ounces of the mushrooms. If you want some variety, use any other sort of mushrooms (white or shiitake, for example) for the other 8 ounces.

Makes 4 servings

INGREDIENTS

- 4 ounces baby bella mushrooms
- 8 ounces mixed other mushrooms or baby bellas
- 12 ounces Swiss chard
- 6 tablespoons unsalted butter
- 4 large garlic cloves, chopped
- 8 ounces small whole-wheat pasta
- 3½ cups chicken broth or vegetable stock
- ½ teaspoon salt
- ½ teaspoon black pepper
- ½ cup grated Parmesan

PREPARATION

1. Wipe the mushroom caps and trim the stems. Cut the baby bella mushrooms into halves; cut the other mushrooms, if used, into pieces about the size of half a baby bella. Trim the stems of the chard and tear the leaves into pieces.

2. In a large saucepan over medium heat, melt 4 tablespoons of the butter, whisking constantly, until the butter begins to bubble. Cook for 5 minutes, or until the butter just begins to brown. Remove from the heat and pour the butter into a bowl.

3. Add the remaining butter to the saucepan over medium heat. When it's melted, add the mushrooms and garlic and cook until the mushrooms soften and take on color, about 6 to 8 minutes. Stir in the pasta, chard, chicken broth, salt, and pepper and bring to a boil. Reduce heat to

medium-low, cover, and simmer, stirring occasionally, until pasta is al dente, about 10 minutes. (It's okay if a little liquid remains in the bottom of the saucepan.)

4. Remove from the heat and stir in the reserved butter and Parmesan.

Walnut Pesto Pasta with Greens

This recipe calls for arugula and spinach as the greens, but you can use just one or the other if you prefer or substitute Swiss chard for the spinach.

Makes 4 servings

INGREDIENTS

- 1 box whole-wheat thin spaghetti
- 3/4 cup walnuts
- 3 garlic cloves
- 2 1/2 cups basil
- 3/4 cup grated Parmesan
- 1/4 cup extra-virgin olive oil
- Salt
- Black pepper
- 3 cups mixed arugula and spinach leaves, torn
- 1 1/2 cups cherry or grape tomatoes, halved

PREPARATION

1. Cook the spaghetti in a large pot of boiling water until it is al dente, about 7 to 8 minutes.
2. While the pasta cooks, make the pesto. Combine the walnuts and garlic in a food processor. Pulse for 30 seconds. Add the basil and Parmesan and pulse for 30 seconds. With the motor running, slowly drizzle in the olive oil and process until smooth. Season to taste with salt and pepper.
3. When pasta is done, drain it in a colander, reserving 1 cup of the cooking water, and immediately return the pasta to the pot. Stir in the arugula and spinach—the heat from the pasta will wilt the greens. Add the halved tomatoes. Fold in the pesto; add some of the reserved cooking water if the mixture is too dry. Season with additional salt and pepper.

Zucchini Noodles with Bean Bolognese

The "noodles" in this dish are spirals or ribbons of zucchini or yellow summer squash. The noodles don't have to be cooked, but I think this dish is better if they are.

Makes 4 servings

INGREDIENTS

- 2 medium zucchini or yellow squash
- 2 tablespoons extra-virgin olive oil
- 1 small onion, chopped
- ½ cup chopped carrot
- ¼ cup chopped celery
- 4 cloves garlic, chopped
- ½ teaspoon salt
- 1 14-ounce can diced tomatoes
- 1 teaspoon dried oregano
- ½ teaspoon hot red pepper flakes
- ¼ cup chopped parsley, divided
- 1 15-ounce can cannellini beans, drained and rinsed
- ½ cup grated Parmesan

PREPARATION

1. Spiralize the zucchini or use a vegetable peeler to slice it into thin ribbons. Bring a large pot of water to a boil.
2. Heat the olive oil in a medium saucepan over medium heat. Add the onion, carrot, celery, garlic, and salt. Cook over medium heat, stirring often, until the carrots are softened, about 10 minutes. Add the diced tomatoes, oregano, red pepper flakes, and parsley. Bring to a simmer and cook, stirring often, until the mixture thickens, about 6 minutes.
3. Add the beans and lower the heat. Cook, stirring often, until the beans are heated through, about 3 minutes.
4. After adding the beans to the sauce, drop the noodles into the boiling water and cook for 3 minutes. Drain well in a colander and divide the noodles among four individual serving bowls.
5. Spoon the sauce over the noodles. Pass the Parmesan at the table.

Meals for Kids

Cauliflower Crust Pizza with Tomatoes and Mozzarella

You can make your own cauliflower crust for this healthier version of traditional pizza, but really, who has time for this when the kids are hungry? Frozen cauliflower crusts are the answer. When tomatoes are in season, substitute seeded chunks for the marinara sauce.

Makes 4 servings

INGREDIENTS

- 1 frozen cauliflower pizza crust
- 1½ cups shredded mozzarella
- ½ cup marinara sauce
- ¼ cup basil leaves, torn
- ½ teaspoon hot red pepper flakes

PREPARATION

1. Preheat the oven to 425°F.
2. Remove the frozen cauliflower crust from the wrapping and place it on a baking sheet. Do not defrost in advance.
3. Sprinkle 1¼ cups of the mozzarella over the crust. Spread the marinara sauce over the cheese all the way to the edge of the crust. Sprinkle with the torn basil leaves and red pepper flakes. Top with the remaining cheese.
4. Put the pizza in the oven and bake until the cheese is golden-brown and the crust is crispy, about 12 to 15 minutes.

Healthy Chicken Nuggets

Kids love chicken nuggets, but most fast-food and frozen versions are full of unhealthy food additives, salt, and bad fats. Try this homemade, healthier version. Time-saving hint: put some broccoli florets in a baking dish, toss with a drizzle of olive oil and a sprinkle of salt, and bake them along with the chicken nuggets. It all comes together in about 40 minutes.

Makes 4 servings

INGREDIENTS

1 pound boneless chicken breasts

½ cup almond meal

½ teaspoon paprika

½ teaspoon garlic powder

½ teaspoon salt

½ teaspoon black pepper

2 eggs

PREPARATION

1. Preheat the oven to 375°F. Place a wire rack on top of a baking sheet.
2. Cut the chicken breasts into strips about ½ inch wide.
3. Combine the almond meal, paprika, garlic powder, salt, and black pepper in a medium mixing bowl.
4. Whisk the eggs in a shallow dish.
5. Dip the chicken strips into the egg wash and then roll them firmly in the almond meal mixture. Lay the coated chicken strips on the wire rack.
6. Bake for 10 minutes, then flip the strips over and bake for 5 to 10 more minutes, or until the strips are browned and crispy.

Healthier Mac and Cheese

The Greek yogurt makes this comfort food recipe a little lighter. Chickpea pasta has about twice the protein and three times the fiber of regular semolina pasta—and it's gluten-free. When cooking chickpea pasta, some foam on the cooking water is normal. When it's done, rinse the pasta before using it in the next step of your recipe.

Makes 4 servings

INGREDIENTS

- 1 pound chickpea elbow macaroni
- 2 tablespoons unsalted butter
- 2 tablespoons flour
- 1 teaspoon salt
- ¼ teaspoon black pepper
- 2 cups milk
- 1½ cups grated Cheddar cheese
- ¾ cup plain nonfat Greek yogurt
- 2 tablespoons breadcrumbs

PREPARATION

1. Preheat the oven to 450°F.
2. Cook the pasta in a large pot of boiling water until it is al dente. Be careful not to overcook. Drain the pasta in a colander and rinse with cool water.
3. In a large saucepan, melt the butter over medium heat. Sprinkle in the flour, salt, and pepper and whisk to form a smooth roux.
4. Whisk in the milk ¼ cup at a time and cook until the mixture is thickened. Don't let it boil.
5. Whisk in the cheese and stir until the cheese is melted and the sauce is thick. Stir in the yogurt. Add the pasta and stir gently until it is evenly coated with the sauce.
6. Pour the pasta mixture into an 8×8-inch baking dish or 2-quart gratin dish and spread evenly. Top with breadcrumbs.
7. Bake for 15 minutes, or until the top and breadcrumbs are golden-brown.

Turkey Burgers

Turkey burgers are a leaner alternative to beef burgers. They tend to be on the bland side, however, so don't hesitate to spice up this basic recipe. Chili powder works well; so does a hot Cajun seasoning blend. You can double this recipe and freeze the extra patties by putting them on a baking sheet, freezing for 2 hours, and then transferring them to an airtight container. Thaw in the refrigerator before using.

Makes 4 servings

INGREDIENTS

1 pound ground turkey
¼ cup breadcrumbs
1 onion, finely chopped
1 scallion, chopped
1 egg
2 teaspoons garlic powder
2 tablespoons dried parsley
½ teaspoon salt
¼ teaspoon black pepper

PREPARATION

1. Combine the ground turkey, breadcrumbs, onion, scallion, egg, garlic powder, parsley, salt, and pepper together in a large bowl. Mix well. Cover the bowl and refrigerate for 1 hour or longer.
2. Preheat the oven to 400°F.
3. Form the chilled mixture into 4 patties. Arrange the patties in a baking dish.
4. Bake for 30 minutes, or until the juices run clear and the burgers are no longer pink in the center.

Snacks

When I advise my patients to add more fiber to their diet, I recommend swapping low-nutrition snacks like cookies and chips for healthier choices. Cooked edamame, nuts of any sort, sunflower and pumpkin seeds, dried fruit such as apricots, raisins, mango, and pineapple (get the no-sugar-added kind), and even dried seaweed are convenient and delicious. You can also try hummus or salsa as a dip with vegetable sticks.

Almond Raisin Energy Bites

My boys love these as a snack before cross-country practice.

Makes 6 bites

INGREDIENTS

- ¾ cup raisins
- 1¼ cups quick oats (don't use instant oats)
- 1 teaspoon cinnamon
- ½ cup organic peanut butter or almond butter
- 1 teaspoon vanilla extract
- ¼ cup honey

PREPARATION

1. In a medium mixing bowl combine the raisins, oats, and cinnamon. Add the nut butter, vanilla extract, and honey and stir to combine. Cover the bowl and let it sit in the refrigerator for an hour.
2. Roll the chilled mixture into balls or form it into bars. In the unlikely event any are left, store them in the refrigerator.

Healthy Microwave Popcorn

Packaged microwave popcorn contains added chemical ingredients that are best left out of your diet. Use this incredibly simple method instead. Top the popcorn with the traditional salt and butter, or try something different: grated Parmesan, grated lemon zest, chili powder, garlic powder, and curry powder are all good choices. For a different take, try unsweetened cocoa powder, ground cinnamon—or both.

INGREDIENTS

¼ cup popcorn kernels

PREPARATION

1. Put the popcorn kernels into a brown paper sandwich bag. Fold the top of the bag over 2 or 3 times and place it, folded side down, in the microwave.
2. Microwave on the popcorn setting until you hear the popping slow.
3. Remove from the microwave, pour the popcorn into a bowl, and season to taste.

Baked Plantain Chips

Plantain chips are a good alternative to potato chips—all the crunchy flavor plus lots of prebiotic fiber. Serve with guacamole or salsa.

Makes about 3 cups

INGREDIENTS

- 2 large green plantains
- 2 tablespoons avocado or grapeseed oil
- ½ teaspoon salt
- Grated lime zest

PREPARATION

1. Preheat the oven to 375°F.
2. Peel the plantains and slice them as thinly as possible (use a mandoline if you have one). Put the slices into a medium mixing bowl and add the oil and salt. Toss gently to coat the slices.
3. Arrange the slices in a single layer on a large nonstick baking sheet. Bake for 18 minutes and check to see how they're doing. If they're not crispy and turning golden-brown, bake for another 5 minutes and check again.
4. Remove from the oven and sprinkle with more salt and the lime zest. Let cool before serving. If there are any left, store in an airtight container.

Spinach Artichoke Dip

What better way to get the fabulous prebiotic fiber of artichokes and spinach than this classic dip, modified to lighten it up. The recipe calls for jarred artichoke hearts, but you can use frozen instead. Just cook them first and let cool before adding them. Serve the dip with vegetable sticks, pita chips, or tortilla chips. It will keep in the refrigerator for three days.

Makes about 3 cups

INGREDIENTS

10 ounces frozen spinach, defrosted and drained

1 8-ounce package reduced-fat cream cheese

1 cup plain nonfat Greek yogurt

½ cup jarred artichoke hearts, drained and chopped

¼ cup grated Parmesan

½ cup shredded mozzarella

3 large garlic cloves, finely chopped

¼ teaspoon hot red pepper flakes

1 teaspoon lemon juice

½ teaspoon kosher salt

PREPARATION

1. Preheat the oven to 350°F. Grease a 1-quart baking dish.
2. Squeeze the defrosted spinach in a cheesecloth or a clean kitchen towel to remove excess water.
3. In a large mixing bowl, mix the cream cheese and yogurt together until smooth.
4. Add the spinach, artichokes, Parmesan, mozzarella, garlic, red pepper flakes, lemon juice, and salt and stir well until all the ingredients are well mixed.
5. Pour the mixture into the greased baking dish and bake for 20 to 25 minutes, or until the mixture is lightly browned on top.

Simple Guacamole

Guacamole with plantain chips or vegetable sticks makes a quick, energy-producing snack, so I like to keep some on hand in the fridge. To keep the guacamole from turning brown on top, put it in an airtight storage container and smooth down the surface. Gently pour in just enough cold water to cover the surface by about a quarter inch. Cover and store in the refrigerator. Pour off the water and give the guac a stir before using.

Makes between 2½ and 3 cups, depending on the size of the avocados

INGREDIENTS

- 3 medium-ripe avocados
- ¼ cup finely chopped red onion
- 1 jalapeño pepper, seeded and finely chopped
- ¼ cup finely chopped cilantro leaves and stems
- ½ teaspoon kosher salt
- 2 tablespoons lime juice

PREPARATION

Cut the avocados in half and remove the pits. Scoop out the flesh into a medium mixing bowl. Add the onion, jalapeño, cilantro, and salt. Mix gently to combine; leave some lumps in the avocado. Stir in the lime juice.

Hummus

I always have some convenient canned chickpeas in my pantry—they're a great source of prebiotic fiber. Hummus is a wonderful snack or quick lunch with vegetable sticks or toasted pita bread. Try it as an alternative to mayo on a sandwich. When preparing this quick version, be sure to take the tahini out of the fridge in advance and let it come to room temperature before you try to measure it out.

Makes about 2 cups

INGREDIENTS

- 1 15-ounce can chickpeas, drained and rinsed
- 6 tablespoons tahini
- 6 tablespoons water
- 2 tablespoons lemon juice
- 1 large garlic clove, finely chopped
- 1 teaspoon kosher salt
- ½ teaspoon cumin

PREPARATION

1. Put the chickpeas into a gallon-size zip-top plastic bag. Lay the bag flat on the counter and use a rolling pin to crush the chickpeas into coarse pieces. Alternatively, use a food processor to coarsely chop the chickpeas.
2. Whisk together the tahini, water, lemon juice, garlic, salt, and cumin in a large mixing bowl. Add the chickpeas and stir until all the ingredients are well combined. Store in an airtight container in the refrigerator for up to a week.

Smoothies

Post-Workout Protein Smoothie

When I have the time after exercising, I like to make myself a protein smoothie. It hydrates me, tastes great, and gives me an energy boost—plus it's a healthy reward for exercising. Once you've mastered this basic recipe (it doesn't take long), have fun with it and invent your own blends. Add or substitute ingredients such as a tablespoon or two of chia, hemp, flax or pumpkin seeds, coconut milk, unsweetened coconut flakes, or some chopped dates. For an extra boost, add some chopped raw kale or spinach. And definitely try adding a tablespoon of cocoa nibs or unsweetened cocoa powder.

Makes 1 serving

INGREDIENTS

1 cup frozen fruit or berries or 1 ripe banana

3 tablespoons peanut butter or almond butter

½ cup plain nonfat Greek yogurt

½ to 1 cup water, nut milk, oat milk, or orange juice

1 tablespoon honey (optional)

PREPARATION

Combine everything in a blender and blend on high for 1 minute or until smooth. Add more water or an ice cube or two if the smoothie isn't blending well or is too thick for your taste.

Skin So Smoothie

This really simple, fast smoothie is great for the skin. The fruit gives you antioxidants, the yogurt gives you probiotics, and the wheat germ provides a vitamin E boost. I like this for a quick breakfast when I'm having a rushed morning.

Makes 1 serving

INGREDIENTS

- 1 cup plain nonfat Greek yogurt
- 1 cup fresh or frozen blueberries
- ½ cup fresh or frozen mango chunks
- 1 tablespoon wheat germ
- 3 ice cubes

PREPARATION

Combine all the ingredients in a blender and process until smooth.

Desserts

Healthy Frozen Yogurt

Healthy dessert in under 10 minutes. Any kind of frozen fruit, singly or in combination, works.

Makes 4 servings

INGREDIENTS

4 cups frozen fruit

½ cup plain nonfat Greek yogurt

2 teaspoons vanilla extract

3 tablespoons honey

PREPARATION

1. Combine all ingredients in the bowl of a food processor and process until the mixture is smooth and creamy, about 5 minutes or less, depending on the type and size of the frozen fruit.
2. Serve immediately or store in the freezer in an airtight container. Defrost before serving.

Blueberry Cobbler

I always have a bag of blueberries in the freezer, but you can use other fresh or frozen berries to make this easy dessert.

Makes 4 to 6 servings

INGREDIENTS

Fruit base

4 cups fresh or frozen blueberries

2 tablespoons maple syrup

1 tablespoon lemon juice

1 teaspoon vanilla extract

1 tablespoon cornstarch

Cobbler topping

1 cup old-fashioned oats (do not use instant oats)

1 cup chopped walnuts, almonds, or pecans

1 cup almond flour

½ teaspoon salt

½ cup maple syrup

⅓ cup walnut or grapeseed oil

1 teaspoon vanilla extract

PREPARATION

1. Preheat the oven to 350°F.
2. In a large mixing bowl, combine the blueberries, maple syrup, lemon juice, vanilla, and cornstarch. Toss to combine.
3. Spoon the berry mixture into an 8×8-inch baking dish. Leave out any juices still in the bowl.
4. In another mixing bowl, combine the oats, nuts, almond flour, and salt. Stir to combine, then add maple syrup, oil, and vanilla. Stir well.
5. Spread the cobbler topping over the berries; don't worry if there are some gaps. Bake for 40 to 45 minutes, or until the topping is golden-brown. Let cool for 15 minutes before serving.

Chocolate Mousse

Quick, easy, chocolatey—my favorite homemade dessert.

Makes 4 servings

INGREDIENTS

- 3/4 cup milk or milk substitute
- 3 ounces dark chocolate, chopped
- 1 tablespoon honey
- 1/2 teaspoon vanilla extract
- 2 cups plain nonfat Greek yogurt

PREPARATION

1. Combine the milk and chocolate in a small saucepan over medium heat. Stir until the chocolate melts, then stir in the honey and vanilla extract.
2. Put the yogurt in a medium mixing bowl and slowly pour in the chocolate mixture. Stir until evenly blended. Cover the bowl and chill in the refrigerator for 2 hours.
3. Serve cold, topped with berries or other fruit.

Baked Bananas

I call this dessert, but I've been known to eat it for breakfast with some low-sugar granola. The recipe below is for just one serving. To make more, just multiply the ingredients by whatever number of servings you want. For a nutrition boost, sprinkle the banana with chopped walnuts or sliced almonds before you put it into the oven.

Makes 1 serving

INGREDIENTS

- 1 medium-ripe banana, cut in half lengthwise
- ½ tablespoon honey
- ¼ teaspoon ground cinnamon, nutmeg, or allspice

PREPARATION

1. Preheat the oven to 400°F.
2. Put the banana halves in a baking dish with a cover. Drizzle with the honey and sprinkle with the spice.
3. Cover the dish and bake for 12 minutes.

Fruit Compote

A simple fruit topping that's great on pancakes. I like mango and blueberries, but any combination will work. If you use fresh fruit, cut it into pieces about the size of a strawberry.

Makes 3 cups

INGREDIENTS

3 cups fresh or frozen fruit

3 tablespoons orange juice

¼ teaspoon ground cinnamon, nutmeg, or allspice

¼ teaspoon ground ginger

PREPARATION

1. Combine the fruit and the orange juice in a medium saucepan over medium heat. Cook until the mixture is bubbling, then reduce the heat to a simmer and cook, stirring often to break up the fruit, for 10 to 12 minutes.
2. Remove from the heat and stir in the spice and ginger. Serve warm.
3. Store leftovers in an airtight container in the refrigerator or freeze them in ice cube trays. Reheat before serving.

ACKNOWLEDGMENTS

First, let me thank Sheila Buff for her fantastic research and writing, and also for working with me at lightning speed and putting up with my silly puns.

I must also thank my agent and friend, Lisa Leshne. When we first met a few years ago, I had the kernel of the idea for this book, but life became complicated, and I had to shelve the project. Lisa never gave up on me. Thanks to her incredible support and brilliant guidance, I was motivated to attack this project with renewed energy and passion. The entire Leshne Agency was instrumental in making this book a reality.

The full team at HarperCollins deserves a special thanks, but especially Anna Montague and Lisa Sharkey. Editing with Anna was such a pleasure, and I am thrilled that Lisa's legendary talents have helped shape the book from day one.

I want to thank my patients, who honor me by letting me into their intimate lives and who teach me lessons every day.

And a final thank you to Sandy, for just being there.

NOTES

Chapter I: Meet Your General Contractor: Your Microbiome

1. Wilmanski T, Diener C, Rappaport N, et al. Gut Microbiome Pattern Reflects Healthy Ageing and Predicts Survival in Humans. Nat Metab. 2021 Feb;3(2):274–286. doi: 10.1038/s42255-021-00348–0. Epub 2021 Feb 18. PMID: 33619379; PMCID: PMC8169080. Erratum in: Nat Metab. 2021 Apr;3(4):586.
2. Yeoh YK, Zuo T, Lui GC, et al. Gut Microbiota Composition Reflects Disease Severity and Dysfunctional Immune Responses in Patients with COVID-19. Gut. 2021 Apr;70(4):698–706. Published Online First: 11 January 2021. doi: 10.1136/gutjnl-2020–323020.
3. Elshazli RM, Kline A, Elgaml A, et al. Gastroenterology Manifestations and COVID-19 Outcomes: A Meta-analysis of 25,252 Cohorts among the First and Second Waves. J Med Virol. 2021 May;93(5):2740–2768. doi: 10.1002/jmv.26836. Epub 2021 Feb 23. PMID: 33527440; PMCID: PMC8014082.
4. Louca P, Murray B, Klaser K, et al. Modest Effects of Dietary Supplements During the COVID-19 Pandemic: Insights from 445 850 Users of the COVID-19 Symptom Study App. BMJ Nutr Prev Health 2021;4(1): 149–157. doi: 10.1136/bmjnph-2021–000250.
5. Blaabjerg S, Artzi DM, Aabenhus R. Probiotics for the Prevention of Antibiotic-Associated Diarrhea in Outpatients—A Systematic Review and Meta-analysis. Antibiotics (Basel). 2017;6(4).pii:E21.
6. Ford AC, Harris LA, Lacy BE, et al. Systematic Review with Meta-analysis: The Efficacy of Prebiotics, Probiotics, Synbiotics and Antibiotics in Irritable Bowel Syndrome. Aliment Pharmacol Thera. 2018;48(10):1044–1060.
7. Gadelha CJMU, Bezerra AN. Effects of Probiotics on the Lipid Profile: Systematic Review. J Vasc Bras. 2019;18:e20180124. Published 2019 Aug 9. doi: 10.1590/1677–5449.180124.
8. Davis LMG, Martínez I, Walter J, Hutkins R. A Dose Dependent Impact of Prebiotic Galactooligosaccharides on the Intestinal Microbiota of Healthy Adults. Int J Food Microbiol. 2010 Dec 15;144(2):285–292. doi: 10.1016/j.ijfoodmicro.2010.10.007.

Chapter 2: The Architect: Your Brain

1. Yano JM, Yu K, Donaldson GP, et al. Indigenous Bacteria from the Gut Microbiota Regulate Host Serotonin Biosynthesis. Cell. 2015 Apr 9;161(2):264–276. doi: 10.1016/j.cell.2015.02.047. PMID: 25860609; PMCID: PMC4393509. Erratum in: Cell. 2015 Sep 24;163:258.
2. Parker A, Fonseca S, Carding SR. Gut Microbes and Metabolites as Modulators of Blood-Brain Barrier Integrity and Brain Health. Gut Microbes. 2020;11(2):135–157. doi: 10.1080/19490976.2019.1638722.
3. Yang J, Zheng P, Li Y, et al. Landscapes of Bacterial and Metabolic Signatures and Their Interaction in Major Depressive Disorders. Sci Adv. 2020 Dec 2;6(49):eaba8555. doi: 10.1126/sciadv.aba8555. PMID: 33268363; PMCID: PMC7710361.
4. Martami F, Togha M, Seifishahpar M, et al. The Effects of a Multispecies Probiotic Supplement on Inflammatory Markers and Episodic and Chronic Migraine Characteristics: A Randomized Double-Blind Controlled Trial. Cephalalgia. 2019 Jun;39(7):841–853. doi: 10.1177/0333102418820102. Epub 2019 Jan 8. PMID: 30621517.
5. Guo M, Peng J, Huang X, et al. Gut Microbiome Features of Chinese Patients Newly Diagnosed with Alzheimer's Disease or Mild Cognitive Impairment. J Alzheimers Dis. 2021;80(1):299–310. doi: 10.3233/JAD-201040. PMID: 33523001.
6. Vogt NM, Kerby RL, Dill-McFarland KA, et al. Gut Microbiome Alterations in Alzheimer's Disease. Sci Rep. 2017 Oct 19;7(1):13537. doi: 10.1038/s41598-017-13601-y. PMID: 29051531; PMCID: PMC5648830.
7. Zhang R, Miller RG, Gascon R, et al. Circulating Endotoxin and Systemic Immune Activation in Sporadic Amyotrophic Lateral Sclerosis (sALS). J Neuroimmunol. 2009;206:121–124.
8. Armstrong NM, Tom SE, Harrati A, et al. Longitudinal Relationship of Leisure Activity Engagement with Cognitive Performance Among Non-Demented, Community-Dwelling Older Adults. Gerontologist. 2021 Mar 30:gnab046. doi: 10.1093/geront/gnab046. Epub ahead of print. PMID: 33784376.
9. Holt-Lunstad J, Smith TB, Layton JB. Social Relationships and Mortality Risk: A Meta-analytic Review. PLoS Med. 2010 Jul 27;7(7):e1000316. doi: 10.1371/journal.pmed.1000316. PMID: 20668659; PMCID: PMC2910600.
10. Donovan NJ, Blazer D. Social Isolation and Loneliness in Older Adults: Review and Commentary of a National Academies Report. Am J Geriatr Psychiatry. 2020 Dec;28(12):1233–1244. doi: 10.1016/j.jagp.2020.08.005. Epub 2020 Aug 19. PMID: 32919873; PMCID: PMC7437541.
11. Sommerlad A, Ruegger J, Singh-Manoux A, et al. Marriage and Risk of Dementia: Systematic Review and Meta-analysis of Observational Studies. J Neurol Neurosurg Psychiatry. 2018;89:231–238.
12. Dill-McFarland KA, Tang ZZ, Kemis JH, et al. Close Social Relationships Correlate with Human Gut Microbiota Composition. Sci Rep. 2019 Jan 24;9(1):703. doi: 10.1038/s41598-018-37298–9. PMID: 30679677; PMCID: PMC6345772.

13. Park SQ, Kahnt T, Dogan A, et al. A Neural Link Between Generosity and Happiness. Nat Commun. 2017 Jul 11;8:15964. doi: 10.1038/ncomms15964. PMID: 28696410; PMCID: PMC5508200.

Chapter 3: The Kitchen: Eating Your Way to a Younger Body and Mind

1. Eckel RH, Jakicic JM, Ard JD, et al. 2013 AHA/ACC Guideline on Lifestyle Management to Reduce Cardiovascular Risk: A Report of the American College of Cardiology / American Heart Association Task Force on Practice Guidelines. J Am Coll Cardiol. 2014;63(25 Pt B):2960–2984. PMID: 24239922.
2. Ballarini T, et al. Mediterranean Diet, Alzheimer Disease Biomarkers and Brain Atrophy in Old Age. Neurology. 2021 May 5;96(24):e2920–e2922. doi: 10.1212/WNL.0000000000012067. Epub ahead of print. PMID: 33952652.
3. Hahn VS, Knutsdottir H, Luo X, et al. Myocardial Gene Expression Signatures in Human Heart Failure with Preserved Ejection Fraction. Circulation. 2021 Jan 12;143(2):120–134. doi: 10.1161/CIRCULATIONAHA.120.050498. Epub 2020 Oct 29. PMID: 33118835; PMCID: PMC7856095.
4. Morris MC, Tangney CC, Wang Y, et al. MIND Diet Slows Cognitive Decline with Aging. Alzheimers Dement. 2015;11(9):1015–1022. doi:10.1016/j.jalz.2015.04.011.
5. Morris MC, Tangney CC, Wang Y, et al. MIND Diet Associated with Reduced Incidence of Alzheimer's Disease. Alzheimers Dement. 2015 Sep;11(9):1007–1014. doi: 10.1016/j.jalz.2014.11.009. Epub 2015 Feb 11. PMID: 25681666; PMCID: PMC4532650.
6. Asnicar F, Berry SE, Valdes AM, et al. Microbiome Connections with Host Metabolism and Habitual Diet from 1,098 Deeply Phenotyped Individuals. Nat Med. 2021 Feb;27(2):321–332. doi: 10.1038/s41591-020-01183–8. Epub 2021 Jan 11. PMID: 33432175.
7. David LA, Maurice CF, Carmody RN, et al. Diet Rapidly and Reproducibly Alters the Human Gut Microbiome. Nature. 2014 Jan 23;505(7484):559–563. doi: 10.1038/nature12820. Epub 2013 Dec 11. PMID: 24336217; PMCID: PMC3957428.
8. Wang Z, Bergeron N, Levison BS, et al. Impact of Chronic Dietary Red Meat, White Meat, or Non-meat Protein on Trimethylamine N-oxide Metabolism and Renal Excretion in Healthy Men and Women. Eur Heart J. 2019 Feb 14;40(7):583–594. doi: 10.1093/eurheartj/ehy799.
9. Ghosh TS, Rampelli S, Jeffery IB, et al. Mediterranean Diet Intervention Alters the gut Microbiome in Older People Reducing Frailty and Improving Health Status: The NU-AGE 1-Year Dietary Intervention Across Five European Countries. Gut. 2020 Jul;69(7):1218–1228. doi: 10.1136/gutjnl-2019–319654. Epub 2020 Feb 17. PMID: 32066625; PMCID: PMC7306987.
10. Chapman MA. The Role of the Colonic Flora in Maintaining a Healthy Large Bowel Mucosa. Ann R Coll Surg Engl. 2001 Mar;83(2):75-80. PMID: 11320933; PMCID: PMC2503330.

11. Ríos-Covián D, Ruas-Madiedo P, Margolles A, et al. Intestinal Short-Chain Fatty Acids and Their Link with Diet and Human Health. Front Microbiol. 2016. doi: 10.3389/fmicb.2016.00185
12. Oliver A, Chase AB, Weihe C, et al. High-Fiber, Whole-Food Dietary Intervention Alters the Human Gut Microbiome but Not Fecal Short-Chain Fatty Acids. mSystems. 2021 Mar 16;6(2):e00115–e00121. doi: 10.1128/mSystems.00115–21. PMID: 33727392
13. Thompson SV, Bailey MA, Taylor AM, et al. Avocado Consumption Alters Gastrointestinal Bacteria Abundance and Microbial Metabolite Concentrations Among Adults with Overweight or Obesity: A Randomized Controlled Trial. J Nutr. 2021 Apr 8;151(4):753-762. doi: 10.1093/jn/nxaa219. PMID: 32805028; PMCID: PMC8030699.
14. Gurwara A, Dai A, Ajami N, et al. Caffeine Consumption and the Colonic Mucosa-Associated Gut Microbiota. Am J Gastroenterol. 2019 Oct;114:S119–S120. doi: 10.14309/01.ajg.0000590316.43252.64.
15. Chen Y, Wu Y, Du M, et al. An Inverse Association Between Tea Consumption and Colorectal Cancer Risk. Oncotarget. 2017 Jun 6;8(23):37367–37376. doi: 10.18632/oncotarget.16959. PMID: 28454102; PMCID: PMC5514915.
16. Peterson CT, Vaughn AR, Sharma V, et al. Effects of Turmeric and Curcumin Dietary Supplementation on Human Gut Microbiota: A Double-Blind, Randomized, Placebo-Controlled Pilot Study. J Evid Based Integr Med. 2018 Jan–Dec;23:2515690X18790725.
17. Yashin A, Yashin Y, Xia X, Nemzer B. Antioxidant Activity of Spices and Their Impact on Human Health: A Review. Antioxidants (Basel). 2017;6(3):70. Published 2017 Sep 15. doi: 10.3390/antiox6030070.
18. Le Roy CI, Wells PM, Si J, et al. Red Wine Consumption Associated with Increased Gut Microbiota α-Diversity in 3 Independent Cohorts. Gastroenterology. 2020 Jan;158(1):270–272.e2. doi: 10.1053/j.gastro.2019.08.024.
19. Yin C, Noratto GD, Fan X, et al. The Impact of Mushroom Polysaccharides on Gut Microbiota and Its Beneficial Effects to Host: A Review. Carbohydr Polym. 2020 Dec 15;250:116942. doi: 10.1016/j.carbpol.2020.116942. Epub 2020 Aug 27. PMID: 33049854. Jayachandran M, Xiao J, Xu B. A Critical Review on Health Promoting Benefits of Edible Mushrooms Through Gut Microbiota. Int J Mol Sci. 2017 Sep 8;18(9):1934. doi: 10.3390/ijms18091934. PMID: 28885559; PMCID: PMC5618583.
20. Suez J, Korem T, Zilberman-Schapira G, et al. Non-caloric Artificial Sweeteners and the Microbiome: Findings and Challenges. Gut Microbes. 2015;6(2):149–155. doi: 10.1080/19490976.2015.1017700. Epub 2015 Apr 1. PMID: 25831243; PMCID: PMC4615743.
21. Saad MJA, Santos A, Prada PO. Linking Gut Microbiota and Inflammation to Obesity and Insulin Resistance. Physiology (Bethesda). 2016 Jul;31(4):283–293. doi: 10.1152/physiol.00041.2015. PMID: 27252163.

22. Turnbaugh PJ, Ley RE, Mahowald MA, et al. An Obesity-Associated Gut Microbiome with Increased Capacity for Energy Harvest. Nature. 2006 Dec 21;444(7122):1027–1031. doi: 10.1038/nature05414. PMID: 17183312.
23. Meslier V, Laiola M, Roager HM, et al. Mediterranean Diet Intervention in Overweight and Obese Subjects Lowers Plasma Cholesterol and Causes Changes in the Gut Microbiome and Metabolome Independently of Energy Intake. Gut. 2020 Jul;69(7):1258–1268. doi: 10.1136/gutjnl-2019–320438. Epub 2020 Feb 19. PMID: 32075887; PMCID: PMC7306983.
24. Alcock J, Maley CC, Aktipis CA. Is Eating Behavior Manipulated by the Gastrointestinal Microbiota? Evolutionary Pressures and Potential Mechanisms. Bioessays. 2014;36(10):940–949. doi: 10.1002/bies.201400071.
25. Savaiano DA, Ritter AJ, Klaenhammer TR, et al. Improving Lactose Digestion and Symptoms of Lactose Intolerance with a Novel Galacto-Oligosaccharide (RP-G28): A Randomized, Double-Blind Clinical Trial. Nutr J. 2013;12:160. doi: 10.1186/1475-2891-12-160.
26. Oak SJ, Jha R. The Effects of Probiotics in Lactose Intolerance: A Systematic Review. Crit Rev Food Sci Nutr. 2019;59(11):1675–1683. doi: 10.1080/10408398.2018.1425977. Epub 2018 Feb 9. PMID: 29425071.
27. Cheng J, Ouwehand AC. Gastroesophageal Reflux Disease and Probiotics: A Systematic Review. Nutrients. 2020;12(1):132. Published 2020 Jan 2. doi: 10.3390/nu12010132.

Chapter 4: The Bathroom: Eliminating Issues

1. American Gastroenterological Association, Bharucha AE, Dorn SD, et al. American Gastroenterological Association Medical Position Statement on Constipation. Gastroenterology. 2013;144(1):211–217.
2. Ohkusa T, Koido S, Nishikawa Y, Sato N. Gut Microbiota and Chronic Constipation: A Review and Update. Front Med (Lausanne). 2019;6:19. Published 2019 Feb 12. doi: 10.3389/fmed.2019.00019.
3. Huaman JW, Mego M, Manichanh C, et al. Effects of Prebiotics vs a Diet Low in FODMAPs in Patients with Functional Gut Disorders. Gastroenterology. 2018 Oct;155(4):1004–1007. doi: 10.1053/j.gastro.2018.06.045. Epub 2018 Jun 30. PMID: 29964041.
4. Parkes GC, Sanderson JD, Whelan K. Treating Irritable Bowel Syndrome with Probiotics: The Evidence. Proc Nutr Soc. 2010 May;69(2):187–194. doi: 10.1017/S002966511000011X. Epub 2010 Mar 18. PMID: 20236566.
5. El-Salhy M, Ystad SO, Mazzawi T, Gundersen D. Dietary Fiber in Irritable Bowel Syndrome (Review). Int J Mol Med. 2017;40(3):607–613. doi: 10.3892/ijmm.2017.3072.
6. Moser G, Trägner S, Gajowniczek EE, et al. Long-Term Success of GUT-Directed Group Hypnosis for Patients with Refractory Irritable Bowel Syndrome: A Randomized Controlled Trial. Am J Gastroenterol. 2013;108(4):602–609.

7. Gaylord SA, Palsson OS, Garland EL, et al. Mindfulness Training Reduces the Severity of Irritable Bowel Syndrome in Women: Results of a Randomized Controlled Trial. Am J Gastroenterol. 2011 Sep;106(9):1678–1688. doi: 10.1038/ajg.2011.184. Epub 2011 Jun 21. PMID: 21691341; PMCID: PMC6502251.
8. Rahimi R, Abdollahi M. Herbal Medicines for the Management of Irritable Bowel Syndrome: A Comprehensive Review. World J Gastroenterol. 2012 Feb 21;18(7):589–600. doi: 10.3748/wjg.v18.i7.589. PMID: 22363129; PMCID: PMC3281215.
9. Parkes GC, Sanderson JD, Whelan K. Treating Irritable Bowel Syndrome with Probiotics: The Evidence. Proc Nutr Soc. 2010 May;69(2):187–194. doi: 10.1017/S002966511000011X. Epub 2010 Mar 18. PMID: 20236566.
10. Tursi A, Papa A, Danese S. Review Article: The Pathophysiology and Medical Management of Diverticulosis and Diverticular Disease of the Colon. Aliment Pharmacol Ther. 2015;42(6):664–684.
11. Perrott S, McDowell R, Murchie P, et al. Global Rise in Early-Onset Colorectal Cancer: An Association with Antibiotic Consumption? Ann. Oncol. 2021 Jul; 32:S213.
12. Veettil SK, Wong TY, Loo YS, et al. Role of Diet in Colorectal Cancer Incidence: Umbrella Review of Meta-analyses of Prospective Observational Studies. JAMA Netw Open. 2021 Feb 1;4(2):e2037341. doi: 10.1001/jamanetworkopen.2020.37341. PMID: 33591366; PMCID: PMC7887658.
13. Modi RM, Hinton A, Pinkhas D, et al. Implementation of a Defecation Posture Modification Device: Impact on Bowel Movement Patterns in Healthy Subjects. J Clin Gastroenterol. 2019 Mar;53(3):216–219. doi: 10.1097/MCG.0000000000001143. PMID: 30346317; PMCID: PMC6382038.

Chapter 5: The Powder Room: Beauty Isn't Just Skin Deep

1. Gkogkolou P, Böhm M. Advanced Glycation End Products: Key Players in Skin Aging? Dermatoendocrinol. 2012;4(3):259–270. doi: 10.4161/derm.22028.
2. de Miranda RB, Weimer P, Rossi RC. Effects of Hydrolyzed Collagen Supplementation on Skin Aging: A Systematic Review and Meta-analysis. Int J Dermatol. 2021 Mar 20. doi: 10.1111/ijd.15518. Epub ahead of print. PMID: 33742704.
3. Salem I, Ramser A, Isham N, Ghannoum MA. The Gut Microbiome as a Major Regulator of the Gut-Skin Axis. Front Microbiol. 2018;9:1459. Published 2018 Jul 10. doi: 10.3389/fmicb.2018.01459.
4. Stahl W, Heinrich U, Aust O, et al. Lycopene-Rich Products and Dietary Photoprotection. Photochem Photobiol Sci. 2006;5(2):238–242. doi: 10.1039/b505312a.
5. Garcia-Peterson LM, Wilking-Busch MJ, Ndiaye MA, et al. Sirtuins in Skin and Skin Cancers. Skin Pharmacol Physiol. 2017;30(4):216–224. doi: 10.1159/000477417.
6. Placzek M, Gaube S, Kerkmann U, et al. Ultraviolet B–Induced DNA Damage in Human Epidermis Is Modified by the Antioxidants Ascorbic Acid and D-alpha-tocopherol. J Invest Dermatol. 2005;124:304–307.

7. Reygagne P, Bastien P, Couavoux MP, et al. The Positive Benefit of Lactobacillus paracasei NCC2461 ST11 in Healthy Volunteers with Moderate to Severe Dandruff. Benef Microbes. 2017 Oct 13;8(5):671–680. doi: 10.3920/BM2016.0144. Epub 2017 Aug 9. PMID: 28789559.
8. Callewaert C, Lambert J, Van de Wiele T. Towards a Bacterial Treatment for Armpit Malodour. Exp Dermatol. 2017;26(5):388–391. doi: 10.1111/exd.13259.
9. Kober MM, Bowe WP. The Effect of Probiotics on Immune Regulation, Acne, and Photoaging. Int J Womens Dermatol. 2015;1(2):85–89. Published 2015 Apr 6. doi: 10.1016/j.ijwd.2015.02.001. Knackstedt R, Knackstedt T, Gatherwright J. The Role of Topical Probiotics in Skin Conditions: A Systematic Review of Animal and Human Studies and Implications for Future Therapies. Exp Dermatol. 2020;29(1):15–21. doi: 10.1111/exd.14032. Yu Y, Dunaway S, Champer J, et al. Changing Our Microbiome: Probiotics in Dermatology. Br J Dermatol. 2020;182(1):39–46. doi: 10.1111/bjd.18088.
10. Janvier X, Alexandre S, Boukerb A, et al. Deleterious Effects of an Air Pollutant (NO2) on a Selection of Commensal Skin Bacterial Strains, Potential Contributor to Dysbiosis? Front Microbiol. 2020 Dec 8;11:591839. doi: 10.3389/fmicb.2020.591839. PMID: 33363523; PMCID: PMC7752777.

Chapter 6: The Home Gym: Work Out to Turn Back the Clock

1. Smith PJ, Blumenthal JA, Hoffman BM, et al. Aerobic Exercise and Neurocognitive Performance: A Meta-analytic Review of Randomized Controlled Trials. Psychosom Med 2010;72:239–252.
2. Mok A, Khaw KT, Luben R, et al. Physical Activity Trajectories and Mortality: Population Based Cohort Study. BMJ. 2019 Jun 26;365:l2323. doi: 10.1136/bmj.l2323. PMID: 31243014; PMCID: PMC6592407.
3. Sallis R, Young DR, Tartof SY, et al. Physical Inactivity Is Associated with a Higher Risk for Severe COVID-19 Outcomes: A Study in 48,440 Adult Patients. Br J Sports Med. 2021 Apr 13:bjsports-2021-104080. doi: 10.1136/bjsports-2021-104080. Epub ahead of print. PMID: 33849909; PMCID: PMC8050880.
4. Zhu Q, Jiang S, Du G. Effects of Exercise Frequency on the Gut Microbiota in Elderly Individuals. Microbiologyopen. 2020 Aug;9(8):e1053. doi: 10.1002/mbo3.1053. Epub 2020 May 1. PMID: 32356611; PMCID: PMC7424259.
5. Hughes RL. A Review of the Role of the Gut Microbiome in Personalized Sports Nutrition. Front Nutr. 2020 Jan 10;6:191. doi: 10.3389/fnut.2019.00191. PMID: 31998739; PMCID: PMC6966970.
6. Scheiman J, Luber JM, Chavkin TA, et al. Meta-omics Analysis of Elite Athletes Identifies a Performance-Enhancing Microbe That Functions via Lactate Metabolism. Nat Med. 2019;25:1104–1109. doi: 10.1038/s41591-019-0485-4.
7. Allen JM, Mailing LJ, Niemiro GM, et al. Exercise Alters Gut Microbiota Composition and Function in Lean and Obese Humans. Med Sci Sports Exerc. 2018 Apr;50(4):747–757. doi: 10.1249/MSS.0000000000001495. PMID: 29166320.

8. Chau JY, Grunseit AC, Chey T, et al. Daily Sitting Time and All-Cause Mortality: A Meta-analysis. PLoS One. 2013;8(11):e80000. Published 2013 Nov 13. doi: 10.1371/journal.pone.0080000.
9. Chau JY, Grunseit AC, Chey T, et al. Daily Sitting Time and All-Cause Mortality: A Meta-analysis. PLoS One. 2013;8(11):e80000. Published 2013 Nov 13. doi: 10.1371/journal.pone.0080000.
10. Patel AV, Maliniak ML, Rees-Punia E, et al. Prolonged Leisure Time Spent Sitting in Relation to Cause-Specific Mortality in a Large US Cohort. Am J Epidemiol. 2018 Oct 1;187(10):2151–2158. doi: 10.1093/aje/kwy125.
11. Jahnke R, Larkey L, Rogers C, et al. A Comprehensive Review of Health Benefits of Qigong and Tai Chi. Am J Health Promot. 2010;24(6):e1–e25.
12. MacEwen BT, MacDonald DJ, Burr JF. A Systematic Review of Standing and Treadmill Desks in the Workplace. Prev Med. 2015 Jan;70:50–58. doi: 10.1016/j.ypmed.2014.11.011. Epub 2014 Nov 28. PMID: 25448843.
13. Bosse JD, Dixon BM. Dietary Protein to Maximize Resistance Training: A Review and Examination of Protein Spread and Change Theories. J Int Soc Sports Nutr. 2012 Sep 8;9(1):42. doi: 10.1186/1550-2783-9-42. PMID: 22958314; PMCID: PMC3518828.
14. Buigues C, Fernández-Garrido J, Pruimboom L, et al. Effect of a Prebiotic Formulation on Frailty Syndrome: A Randomized, Double-Blind Clinical Trial. Int J Mol Sci. 2016 Jun 14;17(6):932. doi: 10.3390/ijms17060932. PMID: 27314331; PMCID: PMC4926465.
15. Nilsson AG, Sundh D, Bäckhed F, Lorentzon M. Lactobacillus reuteri Reduces Bone Loss in Older Women with Low Bone Mineral Density: A Randomized, Placebo-Controlled, Double-Blind, Clinical Trial. J Intern Med. 2018 Sep;284(3):307–317. doi: 10.1111/joim.12805.
16. de Sire A, de Sire R, Petito V, et al. Gut-Joint Axis: The Role of Physical Exercise on Gut Microbiota Modulation in Older People with Osteoarthritis. Nutrients. 2020;12(2):574. Published 2020 Feb 22. doi: 10.3390/nu12020574.
17. Bodkhe R, Balakrishnan B, Taneja V. The Role of Microbiome in Rheumatoid Arthritis Treatment. Ther Adv Musculoskelet Dis. 2019;11:1759720X19844632. Published 2019 Jul 30. doi: 10.1177/1759720X19844632.

Chapter 7: The Zen Corner

1. Johnstone N, Milesi C, Burn O, et al. Anxiolytic Effects of a Galacto-Oligosaccharides Prebiotic in Healthy Females (18–25 Years) with Corresponding Changes in Gut Bacterial Composition. Sci Rep. 2021 Apr 15;11(1):8302. doi: 10.1038/s41598-021-87865-w. PMID: 33859330; PMCID: PMC8050281.
2. Jim Cornall, "New Research Shows FrieslandCampina Ingredients' Biotis GOS Reduces Anxiety," DairyReporter.com, May 20, 2021, https://www.dairyreporter.com/Article/2021/05/20/New-research-shows-FrieslandCampina-Ingredients-Biotis-GOS-reduces-anxiety.

3. Kiecolt-Glaser JK, Wilson SJ, Bailey ML, et al. Marital Distress, Depression, and a Leaky Gut: Translocation of Bacterial Endotoxin as a Pathway to Inflammation. Psychoneuroendocrinology. 2018 Dec;98:52–60. doi: 10.1016/j.psyneuen.2018.08.007. Epub 2018 Aug 4. PMID: 30098513; PMCID: PMC6260591.
4. Goldstein P, Weissman-Fogel I, Dumas G, Shamay-Tsoory SG. Brain-to-Brain Coupling During Handholding Is Associated with Pain Reduction. Proc Natl Acad Sci USA. 2018 Mar 13;115(11):E2528–E2537. doi: 10.1073/pnas.1703643115. Epub 2018 Feb 26. PMID: 29483250; PMCID: PMC5856497.
5. Hunt MG, Marx R, Lipson C, Young J. No More FOMO: Limiting Social Media Decreases Loneliness and Depression. J Soc Clin Psychol. 2018;37(10):751–768.
6. 2020 Stress in America™ survey; the Harris Poll on behalf of the American Psychological Association.
7. Househam AM, Peterson CT, Mills PJ, Chopra D. The Effects of Stress and Meditation on the Immune System, Human Microbiota, and Epigenetics. Adv Mind Body Med. 2017 Fall;31(4):10–25. PMID: 29306937.
8. Gaylord SA, Palsson OS, Garland EL, et al. Mindfulness Training Reduces The Severity of Irritable Bowel Syndrome in Women: Results of a Randomized Controlled Trial. Am J Gastroenterol. 2011 Sep;106(9):1678–1688. doi: 10.1038/ajg.2011.184. Epub 2011 Jun 21. PMID: 21691341; PMCID: PMC6502251.
9. Lambert NM, Fincham FD, Stillman TF. Gratitude and Depressive Symptoms: The Role of Positive Reframing and Positive Emotion. Cogn Emot. 2012;26(4):615–633. doi: 10.1080/02699931.2011.595393. Epub 2011 Sep 19. PMID: 21923564.
10. Stier-Jarmer M, Throner V, Kirschneck M, et al. The Psychological and Physical Effects of Forests on Human Health: A Systematic Review of Systematic Reviews and Meta-Analyses. Int J Environ Res Public Health. 2021 Feb 11;18(4):1770. doi: 10.3390/ijerph18041770. PMID: 33670337; PMCID: PMC7918603.
11. Lackner JM, Jaccard J, Radziwon CD, et al. Durability and Decay of Treatment Benefit of Cognitive Behavioral Therapy for Irritable Bowel Syndrome: 12-Month Follow-up. Am J Gastroenterol. 2019;114:330–338.
12. Martin FP, Montoliu I, Nagy K, et al. Specific Dietary Preferences Are Linked to Differing Gut Microbial Metabolic Activity in Response to Dark Chocolate Intake. J Proteome Res. 2012 Dec 7;11(12):6252–6263. doi: 10.1021/pr300915z. Epub 2012 Nov 19. PMID: 23163751.
13. Wiese M, Bashmakov Y, Chalyk N, et al. Prebiotic Effect of Lycopene and Dark Chocolate on Gut Microbiome with Systemic Changes in Liver Metabolism, Skeletal Muscles and Skin in Moderately Obese Persons. Biomed Res Int. 2019 Jun 2;2019:4625279. doi: 10.1155/2019/4625279. PMID: 31317029; PMCID: PMC6604498. Tuohy KM, Conterno L, Gasperotti M, Viola R. Up-Regulating the Human Intestinal Microbiome Using Whole Plant Foods, Polyphenols, and/or Fiber. J Agric Food Chem. 2012 Sep 12;60(36):8776–8782. doi: 10.1021/jf2053959. Epub 2012 Jun 12. PMID: 22607578.

Chapter 8: The Bedroom: Sleeping Your Way to Better Health

1. "America's State of Mind Report." Express Scripts, April 16, 2020. https://express-scripts.com/corporate/americas-state-of-mind-report.
2. Centers for Disease Control, National Center for Chronic Disease Prevention and Health Promotion, Division of Population Health, Sleep and Sleep Disorders, https://www.cdc.gov/sleep/index.html.
3. Huang T, Redline S. Cross-Sectional and Prospective Associations of Actigraphy-Assessed Sleep Regularity with Metabolic Abnormalities: The Multi-Ethnic Study of Atherosclerosis. Diabetes Care. 2019 Aug;42(8):1422–1429. doi: 10.2337/dc19-0596.
4. Robbins R, Quan SF, Weaver MD, et al. Examining Sleep Deficiency and Disturbance and Their Risk for Incident Dementia and All-Cause Mortality in Older Adults Across 5 Years in the United States. Aging (Albany NY). 2021;13:3254–3268. doi: 10.18632/aging.202591.
5. Fultz NE, Bonmassar G, Setsompop K, et al. Coupled Electrophysiological, Hemodynamic, and Cerebrospinal Fluid Oscillations in Human Sleep. Science. 2019 Nov 1;366(6465):628–631. doi: 10.1126/science.aax5440. PMID: 31672896; PMCID: PMC7309589.
6. St-Onge MP. Sleep-Obesity Relation: Underlying Mechanisms and Consequences for Treatment. Obes Rev. 2017 Feb;18 Suppl 1:34–39. doi: 10.1111/obr.12499. PMID: 28164452.
7. Greer SM, Goldstein AN, Walker MP. The Impact of Sleep Deprivation on Food Desire in the Human Brain. Nat Commun. 2013;4:2259. doi: 10.1038/ncomms3259. PMID: 23922121; PMCID: PMC3763921.
8. West NP, Hughes L, Ramsey R, et al. Probiotics, Anticipation Stress, and the Acute Immune Response to Night Shift. Front Immunol. 2021 Jan 28;11:599547. doi: 10.3389/fimmu.2020.599547. PMID: 33584665; PMCID: PMC7877220.
9. Li Y, Hao Y, Fan F, Zhang B. The Role of Microbiome in Insomnia, Circadian Disturbance and Depression. Front Psychiatry. 2018 Dec 5;9:669. doi: 10.3389/fpsyt.2018.00669. PMID: 30568608; PMCID: PMC6290721.
10. Takada M, Nishida K, Gondo Y, et al. Beneficial Effects of Lactobacillus casei Strain Shirota on Academic Stress-Induced Sleep Disturbance in Healthy Adults: A Double-Blind, Randomised, Placebo-Controlled Trial. Benef Microbes. 2017 Apr 26;8(2):153–162. doi: 10.3920/BM2016.0150. PMID: 28443383. Marotta A, Sarno E, Del Casale A, et al. Effects of Probiotics on Cognitive Reactivity, Mood, and Sleep Quality. Front Psychiatry. 2019;10:164. Published 2019 Mar 27. doi: 10.3389/fpsyt.2019.00164.
11. van Herwaarden MA, Katzka DA, Smout AJ, et al. Effect of Different Recumbent Positions on Postprandial Gastroesophageal Reflux in Normal Subjects. Am J Gastroenterol. 2000 Oct;95(10):2731–2736. doi: 10.1111/j.1572-0241.2000.03180.x. PMID: 11051341.

12. Duboc H, Coffin B, Siproudhis L. Disruption of Circadian Rhythms and Gut Motility: An Overview of Underlying Mechanisms and Associated Pathologies. J Clin Gastroenterol. 2020 May–Jun;54(5):405–414. doi: 10.1097/MCG.0000000000001333. PMID: 32134798; PMCID: PMC7147411.
13. Taylor DJ, Mallory LJ, Lichstein KL, et al. Comorbidity of Chronic Insomnia with Medical Problems. Sleep. 2007 Feb;30(2):213–218. doi: 10.1093/sleep/30.2.213. PMID: 17326547. Erratum in: Sleep. 2007 Jul 1;30(7):table of contents.
14. Irish LA, Kline CE, Gunn HE, et al. The Role of Sleep Hygiene in Promoting Public Health: A Review of Empirical Evidence. Sleep Med Rev. 2015 Aug;22:23–36. doi: 10.1016/j.smrv.2014.10.001. Epub 2014 Oct 16. PMID: 25454674; PMCID: PMC4400203.
15. Koelsch S, Fuermetz J, Sack U, et al. Effects of Music Listening on Cortisol Levels and Propofol Consumption During Spinal Anesthesia. Front Psychol. 2011 Apr 5;2:58. doi: 10.3389/fpsyg.2011.00058. PMID: 21716581; PMCID: PMC3110826.
16. Okamoto-Mizuno K, Mizuno K. Effects of Thermal Environment on Sleep and Circadian Rhythm. J Physiol Anthropol. 2012;31(1):14. Published 2012 May 31. doi: 10.1186/1880-6805-31-14.
17. Ko Y, Lee JY. Effects of Feet Warming Using Bed Socks on Sleep Quality and Thermoregulatory Responses in a Cool Environment. J Physiol Anthropol. 2018 Apr 24;37(1):13. doi: 10.1186/s40101-018-0172-z. PMID: 29699592; PMCID: PMC5921564.
18. Lim EY, Lee SY, Shin HS, et al. The Effect of Lactobacillus acidophilus YT1 (MENOLACTO) on Improving Menopausal Symptoms: A Randomized, Double-Blinded, Placebo-Controlled Clinical Trial. J Clin Med. 2020 Jul 9;9(7):2173. doi: 10.3390/jcm9072173. PMID: 32660010; PMCID: PMC7408745.
19. Shin JH, Park YH, Sim M, et al. Serum Level of Sex Steroid Hormone Is Associated with Diversity and Profiles of Human Gut Microbiome. Res Microbiol. 2019 Jun–Aug;170(4–5):192–201. doi: 10.1016/j.resmic.2019.03.003. Epub 2019 Mar 30. PMID: 30940469.
20. Okamoto T, Hatakeyama S, Imai A, et al. The Association Between Gut Microbiome and Erectile Dysfunction: A Community-based Cross-sectional Study in Japan. Int Urol Nephrol. 2020 Aug;52(8):1421–1428. doi: 10.1007/s11255-020-02443-9. Epub 2020 Mar 19. PMID: 32193686.

Chapter 9: The Nursery: Healthy Guts, Healthy Kids

1. García-Velasco JA, Menabrito M, Catalán IB. What Fertility Specialists Should Know About the Vaginal Microbiome: A Review. Reprod Biomed Online. 2017 Jul;35(1):103–112. doi: 10.1016/j.rbmo.2017.04.005. Epub 2017 Apr 19. PMID: 28479120.
2. Silva MSB, Giacobini P. Don't Trust Your Gut: When Gut Microbiota Disrupt Fertility. Cell Metab. 2019 Oct 1;30(4):616–618. doi: 10.1016/j.cmet.2019.09.005. PMID: 31577927.

3. Fox C, Eichelberger K. Maternal Microbiome and Pregnancy Outcomes. Fertil Steril. 2015 Dec;104(6):1358–1363. doi: 10.1016/j.fertnstert.2015.09.037. Epub 2015 Oct 19. PMID: 26493119.
4. Slykerman RF, Hood F, Wickens K, et al. Effect of Lactobacillus rhamnosus HN001 in Pregnancy on Postpartum Symptoms of Depression and Anxiety: A Randomised Double-Blind Placebo-Controlled Trial. EBioMedicine. 2017 Oct;24:159–165. doi: 10.1016/j.ebiom.2017.09.013. Epub 2017 Sep 14. PMID: 28943228; PMCID: PMC5652021.
5. Dominguez-Bello MG, Costello EK, Contreras M, et al. Delivery Mode Shapes the Acquisition and Structure of the Initial Microbiota Across Multiple Body Habitats in Newborns. Proc Natl Acad Sci USA. 2010;107:11971–11975.
6. Sevelsted A, Stokholm J, Bønnelykke K, Bisgaard H. Cesarean Section and Chronic Immune Disorders. Pediatrics 2015;135:e92–e98.
7. Liu D, Shao L, Zhang Y, Kang W. Safety and Efficacy of Lactobacillus for Preventing Necrotizing Enterocolitis in Preterm Infants. Int J Surg. 2020 Apr;76:79–87. doi: 10.1016/j.ijsu.2020.02.031. Epub 2020 Feb 26. PMID: 32109650.
8. Davis EC, Wang M, Donovan SM. The Role of Early Life Nutrition in the Establishment of Gastrointestinal Microbial Composition and Function. Gut Microbes. 2017 Mar 4;8(2):143–171. doi: 10.1080/19490976.2016.1278104. Epub 2017 Jan 9. PMID: 28068209; PMCID: PMC5390825.
9. Rogier EW, Frantz AL, Bruno MEC, et al. Secretory Antibodies in Breast Milk Promote Long-Term Intestinal Homeostasis by Regulating the Gut Microbiota and Host Gene Expression. Proc Natl Acad Sci USA. 2014 Feb 25;111(8):3074–3079. doi: 10.1073/pnas.1315792111. Epub 2014 Feb 3. PMID: 24569806; PMCID: PMC3939878.
10. Borewicz K, Suarez-Diez M, Hechler C, et al. The Effect of Prebiotic Fortified Infant Formulas on Microbiota Composition and Dynamics in Early Life. Sci Rep. 2019 Feb 21;9(1):2434. doi: 10.1038/s41598-018-38268-x. PMID: 30792412; PMCID: PMC6385197.
11. Davis EC, Dinsmoor AM, Wang M, Donovan SM. Microbiome Composition in Pediatric Populations from Birth to Adolescence: Impact of Diet and Prebiotic and Probiotic Interventions. Dig Dis Sci. 2020;65(3):706–722. doi:10.1007/s10620-020-06092-x.
12. O'Brien CE, Meier AK, Cernioglo K, et al. Early Probiotic Supplementation with B. infantis in Breastfed Infants Leads to Persistent Colonization at 1 year. Pediatr Res. 2021 Mar 24. doi: 10.1038/s41390-020-01350–0. Epub ahead of print. PMID: 33762689
13. Stanislawski MA, Dabelea D, Wagner BD, et al. Gut Microbiota in the First 2 Years of Life and the Association with Body Mass Index at Age 12 in a Norwegian Birth Cohort. mBio. 2018;9(5):e01751-18. Published 2018 Oct 23. doi:10.1128/mBio.01751-18. PMID: 30352933.
14. McDade TW, Rutherford J, Adair L, Kuzawa CW. Early Origins of Inflammation: Microbial Exposures in Infancy Predict Lower Levels of C-reactive Protein in Adulthood. Proc Biol Sci. 2010 Apr 7;277(1684):1129–1137. doi: 10.1098/rspb.2009.1795. Epub 2009 Dec 9. PMID: 20007176; PMCID: PMC2842762.

15. Hullegie S, Bruijning-Verhagen P, Uiterwaal CS, et al. First-Year Daycare and Incidence of Acute Gastroenteritis. Pediatrics. 2016 May;137(5):e20153356. doi: 10.1542/peds.2015–3356. PMID: 27244798.
16. Noverr MC, Huffnagle GB. The 'Microflora Hypothesis' of Allergic Diseases. Clin Exp Allergy. 2005 Dec;35(12):1511–1520. doi: 10.1111/j.1365–2222.2005.02379.x. PMID: 16393316.
17. Lin J, Zhang Y, He C, Dai J. Probiotics Supplementation in Children with Asthma: A Systematic Review and Meta-analysis. J Paediatr Child Health. 2018 Sep;54(9):953–961. doi: 10.1111/jpc.14126. Epub 2018 Jul 27. PMID: 30051941.
18. Arrieta MC, Stiemsma LT, Dimitriu PA, et al. Early Infancy Microbial and Metabolic Alterations Affect Risk of Childhood Asthma. Sci Transl Med. 2015 Sep 30;7(307):307ra152. doi: 10.1126/scitranslmed.aab2271. PMID: 26424567.
19. Fiocchi A, Pawankar R, Cuello-Garcia C, et al. World Allergy Organization–McMaster University Guidelines for Allergic Disease Prevention (GLAD-P): Probiotics. World Allergy Organ J. 2015 Jan 27;8(1):4. doi: 10.1186/s40413-015-0055–2. PMID: 25628773; PMCID: PMC4307749.
20. Hesselmar B, Hicke-Roberts A, Lundell AC, et al. Pet-Keeping in Early Life Reduces the Risk of Allergy in a Dose-Dependent Fashion. PLoS One. 2018 Dec 19;13(12):e0208472. doi: 10.1371/journal.pone.0208472. PMID: 30566481; PMCID: PMC6300190.
21. Shmalberg J, Montalbano C, Morelli G, Buckley GJ. A Randomized Double Blinded Placebo-Controlled Clinical Trial of a Probiotic or Metronidazole for Acute Canine Diarrhea. Front Vet Sci. 2019 Jun 4;6:163. doi: 10.3389/fvets.2019.00163. PMID: 31275948; PMCID: PMC6593266.
22. McElhanon BO, McCracken C, Karpen S, Sharp WG. Gastrointestinal Symptoms in Autism Spectrum Disorder: A Meta-analysis. Pediatrics. 2014 May;133(5):872–883. doi: 10.1542/peds.2013–3995. PMID: 24777214.
23. Doshi-Velez F, Avillach P, Palmer N, et al. Prevalence of Inflammatory Bowel Disease Among Patients with Autism Spectrum Disorders. Inflamm Bowel Dis. 2015 Oct;21(10):2281–2288. doi: 10.1097/MIB.0000000000000502. PMID: 26218138.
24. Mayer EA, Padua D, Tillisch K. Altered Brain-Gut Axis in Autism: Comorbidity or Causative Mechanisms? Bioessays. 2014;36:933–939.
25. David MM, Tataru C, Daniels J, et al. Children with Autism and Their Typically Developing Siblings Differ in Amplicon Sequence Variants and Predicted Functions of Stool-Associated Microbes. mSystems 2021;6(2): e00193–e00220.
26. Leyer GJ, Li S, Mubasher ME, et al. Probiotic Effects on Cold and Influenza-Like Symptom Incidence and Duration in Children. Pediatrics. 2009 Aug;124(2):e172–e179. doi: 10.1542/peds.2008–2666. Epub 2009 Jul 27. PMID: 19651563.
27. Lazou Ahrén I, Berggren A, Teixeira C, et al. Evaluation of the Efficacy of Lactobacillus plantarum HEAL9 and Lactobacillus paracasei 8700:2 on Aspects of Common Cold Infections in Children Attending Day Care: A Randomised, Double-blind,

Placebo-controlled Clinical Study. Eur J Nutr. 2020 Feb;59(1):409–417. doi: 10.1007/s00394-019-02137-8. Epub 2019 Nov 16. Erratum in: Eur J Nutr. 2019 Dec 19. PMID: 31734734; PMCID: PMC7000506.

28. Aversa Z, Atkinson EJ, Schafer MJ, et al. Association of Infant Antibiotic Exposure with Childhood Health Outcomes. Mayo Clin Proc. 2021 Jan;96(1):66–77. doi: 10.1016/j.mayocp.2020.07.019. Epub 2020 Nov 16. PMID: 33208243; PMCID: PMC7796951.
29. Korpela K, Salonen A, Virta LJ, et al. Intestinal Microbiome Is Related to Lifetime Antibiotic Use in Finnish Pre-school Children. Nat Commun. 2016 Jan 26;7:10410. doi: 10.1038/ncomms10410. PMID: 26811868; PMCID: PMC4737757.
30. Hersh AL, Jackson MA, Hicks LA, et al. Principles of Judicious Antibiotic Prescribing for Upper Respiratory Tract Infections in Pediatrics. Pediatrics. 2013;132(6):1146–1154.
31. Goldenberg JZ, Lytvyn L, Steurich J, et al. Probiotics for the Prevention of Pediatric Antibiotic-associated Diarrhea. Cochrane Database Syst Rev. 2015 Dec 22;(12):CD004827. doi: 10.1002/14651858.CD004827.pub4. Update in: Cochrane Database Syst Rev. 2019 Apr 30;4:CD004827. PMID: 26695080.
32. Davis EC, Dinsmoor AM, Wang M, Donovan SM. Microbiome Composition in Pediatric Populations from Birth to Adolescence: Impact of Diet and Prebiotic and Probiotic Interventions. Dig Dis Sci. 2020;65(3):706–722. doi: 10.1007/s10620-020-06092-x.
33. Kumbhare SV, Patangia DVV, Patil RH, et al. Factors Influencing the Gut Microbiome in Children: From Infancy to Childhood. J Biosci. 2019 Jun;44(2):49. PMID: 31180062.

Chapter 10: The Laundry Room: Detoxing Your Home

1. Sanidad KZ, Xiao H, Zhang G. Triclosan, a Common Antimicrobial Ingredient, on Gut Microbiota and Gut Health. Gut Microbes. 2019;10(3):434–437. doi: 10.1080/19490976.2018.1546521. Epub 2018 Nov 20. PMID: 30453815; PMCID: PMC6546352.
2. Tun MH, Tun HM, Mahoney JJ, et al. Postnatal Exposure to Household Disinfectants, Infant Gut Microbiota and Subsequent Risk of Overweight in Children. CMAJ. 2018 Sep 17;190(37):E1097–E1107. doi: 10.1503/cmaj.170809. PMID: 30224442; PMCID: PMC6141245. Erratum in: CMAJ. 2018 Nov 12;190(45):E1341.
3. Herr M, Just J, Nikasinovic L, et al. Influence of Host and Environmental Factors on Wheezing Severity in Infants: Findings from the PARIS Birth Cohort. Clin Exp Allergy 2012;42:275–283.
4. Pew Charitable Trusts. Could Efforts to Fight the Coronavirus Lead to Overuse of Antibiotics? Issue brief, March 10, 2021. https://www.pewtrusts.org/en/research-and-analysis/issue-briefs/2021/03/could-efforts-to-fight-the-coronavirus-lead-to-overuse-of-antibiotics.

5. Claus SP, Guillou H, Ellero-Simatos S. The Gut Microbiota: A Major Player in the Toxicity of Environmental Pollutants? NPJ Biofilms Microbiomes. 2016 May 4;2:16003. doi: 10.1038/npjbiofilms.2016.3. PMID: 28721242; PMCID: PMC5515271. Erratum in: NPJ Biofilms Microbiomes. 2017 Jun 22;3:17001.
6. Gardner CM, Hoffman K, Stapleton HM, Gunsch CK. Exposures to Semivolatile Organic Compounds in Indoor Environments and Associations with the Gut Microbiomes of Children. Environ Sci Technol Lett. 2020. doi: 10.1021/acs.estlett.0c00776.
7. Fouladi F, Bailey MJ, Patterson WB, et al. Air Pollution Exposure Is Associated with the Gut Microbiome as Revealed by Shotgun Metagenomic Sequencing. Environ Int. 2020 May;138:105604. doi: 10.1016/j.envint.2020.105604. Epub 2020 Mar 2. PMID: 32135388; PMCID: PMC7181344.
8. Tsiaoussis J, Antoniou MN, Koliarakis I, et al. Effects of Single and Combined Toxic Exposures on the Gut Microbiome: Current Knowledge and Future Directions. Toxicol Lett. 2019 Sep 15;312:72–97. doi: 10.1016/j.toxlet.2019.04.014. Epub 2019 Apr 27. PMID: 31034867.
9. Rogers MAM, Aronoff DM. The Influence of Non-steroidal Anti-inflammatory Drugs on the Gut Microbiome. Clin Microbiol Infect. 2016 Feb;22(2):178.e1–178.e9. doi: 10.1016/j.cmi.2015.10.003. Epub 2015 Oct 16. PMID: 26482265; PMCID: PMC4754147.
10. Pollard MS, Tucker JS, Green HD. Changes in Adult Alcohol Use and Consequences During the COVID-19 Pandemic in the US. JAMA Netw Open. 2020;3(9):e2022942. doi:10.1001/jamanetworkopen.2020.22942.
11. Engen PA, Green SJ, Voigt RM, et al. The Gastrointestinal Microbiome: Alcohol Effects on the Composition of Intestinal Microbiota. Alcohol Res. 2015;37(2):223–236. PMID: 26695747; PMCID: PMC4590619.
12. Lee SH, Yun Y, Kim SJ, et al. Association Between Cigarette Smoking Status and Composition of Gut Microbiota: Population-Based Cross-Sectional Study. J Clin Med. 2018;7(9):282. Published 2018 Sep 14. doi: 10.3390/jcm7090282.

INDEX

Abdominal (belly) fat, 52, 108
Abdominal pain, 114
  autism and, 154
  colon cancer and, 79
  diverticulosis and, 77
  IBS and, 71–72
  lactose tolerance and, 55
  long COVID and, 16
Acetaminophen (Tylenol), 168–69
Acid reflux, 55–56, 135, 154
Acne, 90–91, 94, 184
  aging and, 87
  diet and, 89, 93
  probiotics for, 86, 98
  retinoids for, 93
Acupuncture, 73
ADHD (attention-deficit/hyperactivity disorder), 152
Adrenal glands, 24, 114
Adrenaline, 115
Advanced glycation end products (AGEs), 88
Aerobic exercise. *See* Exercise
Affirmations, 124, 126, 182
Aging, 2, 13–15
  brain and, 30–33
  diet for, 37–38
  exercise for, 102, 103
  skin and, 87–88
Agricultural chemicals, 10, 165, 166, 167, 168
Air pollution, 10, 88, 91, 93, 98, 185
  indoor, 166–67
Air purifiers, 166–67, 171
Air travel and jet lag, 136–37
Alcohol, 8, 10, 169–70. *See also* Red wine
  Alzheimer's disease and, 29
  bedtime routine and, 139
  colon cancer and, 79
  diarrhea and, 68
  happy hours and work, 172
  hot flashes and, 142
  jet lag and, 136
  stress and, 121–22
Allergies, 171. *See also* Food allergies
  children and infants, 148, 151–53, 155, 156, 158
  skin problems and, 89, 91
Allergy remedies, 169
All-purpose cleaner, 162, 171
Alone time, 32
Alzheimer's disease, 14, 29, 30, 36, 38, 131
Ammonia, 162
Anal itching, 75
Ankylosing spondylitis, 111
Antacids, 56, 67, 168
Antibacterial products, 160, 161, 163
Antibiotic resistance, 162–64
Antibiotics, 8, 67, 155–56, 158, 162–63
Antidepressants, 24, 25–26
Antioxidants, 44–45, 50, 91–93
Anxiety, 6, 8, 115, 119. *See also* Stress management
  diet for, 46, 116, 117
  sleep and, 137, 140
Appetite regulation, 23, 41, 118
Apps
  exercise, 34
  food diary, 83
  meditation, 34
  mindfulness, 121, 182
  sleep, 140, 144
  white noise, 144
Architect. *See* Brain
Arm exercises, 194–95

Armpits and bacteria, 95–96
Arthritis, 10, 109–11
Artificial sweeteners, 10, 51, 56, 181
Aspartame (Splenda), 51
Aspirin, 168
Asthma, 148, 151–53, 155
Atopic dermatitis. *See* Eczema
Autism spectrum disorder (ASD), 153–54
Autoimmune diseases, 10, 11, 13, 110
Autonomic nervous system, 22, 114
Avocado oil, 178
Avocados, 42–43, 48
  recipes, 199, 208, 238

Baby blues, 147
Bacteria. *See* Gut bacteria
Bacteroidetes, 52–53
Balance exercises, 105–6, 111, 192
Bathroom renovations, 80–82
Baths, 76, 144
Beano, 71
Beans, 70, 71, 176, 178
  recipes, 207, 209, 229
Beaver fever, 67
Bedroom environment, 140–43
Bedroom noise, 142, 144
Bedroom temperature, 141
Beds, 140–41
  aiding digestion by raising head of, 135
  side sleeping, 135–36
Bedsheets, 141
Bedtime routine, 135, 138–40, 183–84
Belly fat, 52, 108
Beta-carotene, 91, 92–93
Biceps Curls, 194
Bicycle Crunches, 188
Bidets, 81–82
Bifidobacterium, 17–18, 29, 44
Bifidobacterium infantis (B. infantis), 150
Bilirubin, 62
Blackout shades, 141
Black stool, 62–63
Bleach, 162
Bloating, 8, 10, 16, 69–71
Blood-brain barrier (BBB), 26–27
Blood sugar
  diet for, 36
  fiber for, 41–42
  probiotics for, 19
Body odor (BO), 95–96
Bone density, 109
Bones and microbiome, 109
Bowel movements, 61–62, 81, 82
  color of stool, 62–63, 82
  fiber and, 40–41
Brain, 21–34. *See also* Gut–brain axis
  aging and, 30–33
  best foods for, 30–31, 34, 37–38
  depression and microbiome, 27
  gut bacteria and, 21–22
  migraines and microbiome, 27–28
  neurodegenerative disease, 28–30
  sleep and, 131
Brain exercises, 31, 34, 183
Brain fog, 8, 13
Brain health, 30–33
Brain size, 5
BRAT diet, 68–69
Breakfast, recipes, 198–201
Breakfast cereals, 52, 56, 73
Breast cancer, 42
Breastfeeding, 93, 149–50
Breast milk, 148, 149–50
Brown noise, 142
Burping, 69–71
Butyrate, 27, 41, 103
B vitamins, 118

Cancer, 45, 63. *See also* Colon cancer
  diet for, 42, 46, 47–48, 50
  exercise for, 102
  sleep deficiency and, 130
  smoking and, 171
Candle meditation, 123
Cataracts, 45
Catecholamines, 126
Cats, 152–53
Celiac disease, 10, 53–54, 63, 67, 73, 91

Cell junctions, 9
Cesarean sections (C-sections), 148, 149
CFUs (colony forming units), 18
Cheeses, 28, 176
Chewing, 58
Chewing gums, 66, 70
Chicory fiber, 43
Childbirth, 147–48
Children, 145–58
  diet for, 156–57
  dirt and immune system, 150–52
  microbiome and, 151–53, 156–57
  probiotics for, 152, 154–56
  recipes for, 230–33
Chocolate, 28
Cholesterol
  diet and fiber for, 36, 37, 41, 43, 48, 50, 178
  probiotics for, 19
  sleep deficiency and, 130–31
Chronic idiopathic constipation (CIC), 64–65
Chronic inflammation, 2, 6, 10, 13–15
Chronic stress. *See* Stress
Cinnamon, 46, 179
Circadian rhythms, 130, 131–32, 134, 136
Classic Crunches, 188
Cleaning products, 160
  microbiome and, 161–62
  safe, 162, 171
Clostridioides difficile (C. diff), 18, 164
Coal tar, 95
Coconut oil, 48, 178–79
Coffee (caffeine), 45–46, 68, 136, 139
Cognitive behavioral therapy (CBT), 125, 137
Cognitive decline, 21, 28–30, 50
  sleep deficiency and, 130, 131
Colds, 154–55
Colitis, 39, 152, 153
Colon, 40–41
Colon cancer, 42, 46, 48, 157, 171
  screening (colonoscopies), 77–79, 83
Colon cleansing, 65–66
Colonics, 65–66
Colonoscopies, 2, 75, 77–79, 83
Color of stool, 62–63, 82
Constipation, 8, 63–65, 82, 114, 154
  causes of, 63–64
  definition of, 63
  treatment of, 64–65
Core exercises, 188–89, 193
Corn syrup, 51, 56
"Coronosomnia," 129
Cortisol, 24, 114, 115, 118, 126, 140
COVID-19 pandemic, 4, 129
  antibiotics, 163
  drinking during, 169
  exercise for, 102
  gut and, 15–17
  maskne, 97
  social isolation and, 32–33
  weight gain, 4, 52, 121–22
COVID-19 vaccines, 17
Cramps, 10
Crohn's disease, 10, 91, 110–11
Crossword puzzles, 31
Cruciferous vegetables, 70
Crunches, 188–89
Current good manufacturing processes (cGMPs), 133
Cytokines, 12, 13, 26, 27, 110

Daily affirmations, 124, 126, 182
Daily digital detox, 120, 126, 184
Daily gratitude check-in, 182
Dairy, 54–55, 82, 93
Dandruff, 89, 94–95
Dark chocolate, 43, 126, 178
Deep breathing, 119
Defecation. *See* Bowel movements
Dementia, 28–30, 32–33, 131
Depression, 6, 8, 13
  diet for, 50, 116
  microbiome and, 27
  probiotics for, 18
  sleep and, 134–35

Depressive disorder (MDD), 27
Dermis, 86–87, 91
Desserts, recipes, 242–46
Diabetes, 10
  coffee for, 46
  diet for, 36, 38, 50
  exercise for, 102
  sleep deficiency and, 130–31
Diarrhea, 8, 12, 66–69
  antibiotics and, 155–56, 162
  autism and, 154
  causes of, 66–67
  dairy and, 82
  long COVID and, 16
  microbiome and, 69, 134
  of pet dog, 153
  self-help for, 68–69
  stool color and, 62
  stress and, 114, 115
  traveler's, 67–68
Diet, 35–60
  for brain health, 30–31, 34, 37–38
  for children, 156–57
  for colon cancer, 79
  for constipation, 64–65
  for depression, 116
  for diarrhea, 68–69
  fiber. *See* Fiber
  food and mood, 117–18
  Gut Reno eating plan, 174–79
  health fats, 48–49, 178–79
  Mediterranean, 36–37, 39–40, 111
  microbiome and, 38–40
  MIND, 37–38
  new tools for our kitchen, 45–47
  phytonutrients, 44–45
  prebiotics and probiotics. *See* Prebiotics; Probiotics
  protein, 49–50
  recipes. *See* Gut Reno recipes
  for skin health, 88, 91–93
  for sleep, 133–34, 139
  special diets, 53–57
  sugar and, 50–51
Dietary fiber. *See* Fiber
Diet candies, 66, 70
Digestion (digestive system), 61–62
  dysbiosis and digestive symptoms, 8–9
  gut–brain axis, 22–27
  role of bacteria. *See* Gut bacteria
  sleep and, 135–36
Digital detox, 120, 126, 184
Dirt and immune system, 150–52, 159–60
Distanced self-talk, 127
Diverticulitis, 42, 77
Diverticulosis, 76–77
Docusate, 65
Dogs, 152–53
Dopamine, 118, 140
Drowsiness, 132–33, 138
Dysbiosis, 8–9, 10, 28, 29, 114, 136, 151

Ear infections, 155
Eating habits. *See* Diet
Eating plan. *See* Gut Reno eating plan
Eczema, 91, 94, 98
  children and infants, 148, 151, 152–53, 158
  diet and, 89
Einstein, Albert, 32
Elastin, 87–88, 93, 184
Emulsifiers, 10
Enteric nervous system (ENS), 22, 115
Environmental toxins, 8, 10, 98, 164–66
Environmental Working Group (EWG)
  Clean Fifteen list, 167–68
  Dirty Dozen list, 168
  safe cleaning products, 162
EPA Safer Choice, 162
Epidermis, 86–87, 91, 93
Epigallocatechin-3-gallate (EGCG), 46, 93–94
Epinephrine, 118
Erectile dysfunction (ED), 143–44
Escherichia coli (E. coli), 7, 66
Escitalopram (Lexapro), 24

Esophageal sphincter, 55, 135–36
Estrogen, 109, 142, 143
Exercise, 101–12
  benefits of, 102–3
  bones and microbiome, 109
  for brain health, 31, 34, 183
  for children, 157
  for colon cancer, 79
  for constipation, 65
  eating after workout, 60, 181–82
  for the gut, 102–3
  Gut Reno fitness plan, 179–82
  for jet lag, 136–37
  joints and microbiome, 109–11
  muscles and microbiome, 108
  recommended amount of, 101–2, 107
  for sleep, 139
  types of, 105–7
  workout week. *See* Gut Reno workout week
Exercise partners, 111

Falls, 105, 108
Fats, 8, 35, 48–49, 178–79
FDA (Food and Drug Administration), 19, 49, 160, 168
Fecal microbiota transplantation (FMT), 164
Fecal transplants, 164
Fermented foods, 19, 46, 176
Fertilizers, 167
Fiber, 40–43, 59
  benefits of, 41–42
  best foods for, 174–75
  for children, 157, 158
  for constipation, 64, 65
  for diverticulosis, 77
  for hemorrhoids, 76
  for IBS, 73
  recommended amount of, 42–43, 59, 175
  for skin health, 89–90
Fight-or-flight response, 24, 114, 119
Firmicutes, 29, 52
Fish and seafood, 179
  recipes, 203, 218–21
Fish oil, 48, 185
Fitness plan. *See* Gut Reno fitness plan
5-hydroxytryptophan (5-HTP), 25
Flame retardants, 10, 165–66
Fluoxetine (Prozac), 24
FODMAPs diet, 56–57, 71, 73
Food additives, 8, 10, 38
Food allergies, 10, 91, 151, 152, 153
Food cravings, 51, 53, 114, 133, 134
Food diaries, 67, 83
Food poisoning, 7, 66, 69, 153
Food sensitivities, 10, 66, 70, 83
"Forest bathing," 124
Forward Lunges, 190
Freezing produce, 47
Fructooligosaccharides (FOS), 19, 43
Fruit juices, 51
Fruits, 36, 37, 43
  fiber in, 175
  frozen, 47
  organic, 167–68, 172
  phytonutrients in, 177–78
  rainbow of colors, 59, 177–78
  skin health and, 91–93

Galactooligosaccharides, 43
Gas, 8, 10, 69–71
Gastritis, 168
Gastroenteritis, 154
Gastroenterologists, 1–3, 42
Gastroesophageal reflux disease (GERD), 55–56, 134, 135–36, 171
Gastrointestinal cancer, 42
General contractor. *See* Microbiome
Genetic factors, 2, 28, 29, 79, 150
Ghrelin, 23, 118, 133
Giardia, 67
Gluten, 53–54, 66, 67, 70, 73
Glycosaminoglycans (GAGs), 87
Gratitude, 58, 123, 182
Green stool, 62
Green tea, 46, 93–94, 176

Guided imagery, 123
Gut
  COVID-19 and, 15–17
  exercise for, 102–3
  skin conditions and, 94
Gut bacteria, 5–8
  brain and, 21–22
  COVID-19 and, 15–17
  exercise and, 102–3
  sleep deficiency and, 133–35
  stress and, 115–16
Gut barrier, 9–10
Gut–brain axis, 22–27, 115
  bacterial metabolites, 26–27
  hormones, 23–24
  immune system, 26
  neurotransmitters, 23–25
Gut microbiome. *See* Microbiome
Gut Reno beauty regimen, 184–85
Gut Reno eating plan, 174–79
Gut Reno fitness plan, 179–82
Gut Reno mind plan, 182–83
Gut Reno recipes, 197–246
  Almond Raisin Energy Bites, 234
  Avocado Toast, 199
  Baked Bananas, 245
  Baked Plantain Chips, 236
  Baked Zucchini Fries, 214
  Basic Vinaigrette Dressing, 205
  Blueberry Cobbler, 243
  Buckwheat Pancakes, 201
  Carrots with Cumin, 212
  Cauliflower Crust Pizza with Tomatoes and Mozzarella, 230
  Chocolate Mousse, 244
  Corn and Avocado Salad, 208
  Eggplant Pizza, 224
  Fish Kebabs with Fennel, 220
  Fish Tacos with Mango Salsa, 221
  Fruit Compote, 246
  Healthier Mac and Cheese, 232
  Healthy Chicken Nuggets, 231
  Healthy Frozen Yogurt, 242
  Healthy Microwave Popcorn, 235
  Healthy Tuna Salad, 203
  Hummus, 239
  Kimchi Cauliflower Rice, 217
  Lemony Broccoli Rabe with White Beans, 209
  Miso-Glazed Salmon, 218–19
  Overnight Oats, 200
  Pasta with Mushrooms and Swiss Chard, 226–27
  Portobello Sandwich, 204
  Post-Workout Protein Smoothie, 240
  Probiotic Parfait, 198
  Quinoa Salad with Artichokes, White Beans, and Pistachios, 207
  Roasted Turmeric Chickpeas, 210
  Simple Guacamole, 238
  Simple Kimchi, 215–16
  Skin So Smoothie, 241
  Spinach Artichoke Dip, 237
  Stir-Fried Tofu with Cauliflower, 222–23
  Sweet Potatoes and Peppers, 211
  Sweet Potato Fries, 213
  Turkey Burgers, 233
  Turkey Chili, 225
  Vegetable Frittata, 202
  Walnut Pesto Pasta with Greens, 228
  Watermelon, Spinach, and Tomato Salad, 206
  Zucchini Noodles with Bean Bolognese, 229
Gut Reno sleep plan, 183–84
Gut Reno workout week, 187–96
  Monday, 188–89
  Tuesday, 190
  Wednesday, 191
  Thursday, 192
  Friday, 193
  Saturday, 194–95
Gut–skin axis, 85–87
Gym memberships, 111–12
"Gym rats," 103

Hair follicles, 87, 94–95, 97
Hand sanitizers, 67, 96, 150, 161
Hand washing, 7, 17, 96, 161
Happy hour, 172
Hay fever, 148, 152–53
Headaches, 27–28, 182
Head Rotations, 192
Health fats, 48–49, 178–79
Healthy eating. *See* Diet
Healthy fats, 48
Heartburn, 55–56, 135
Heart disease, 6, 10
  coffee for, 46
  diet for, 36, 37, 38, 39, 50
  exercise for, 102, 105
  sleep deficiency and, 130
Heel Taps, 193
Hemorrhoids, 1, 75–76
Herbal medicine, 119
High blood pressure, 36, 130
High-fructose corn syrup, 51, 56
High-intensity interval training (HIIT), 107, 192
High Knees, 193
Hippocrates, 6
Hives, 89
Home gym. *See* Exercise
Home offices, 107
Homeostasis, 131–32
Hormones, 23–24
Hot flashes, 142
Human Microbiome Project, 6
Human milk oligosaccharides (HMOs), 149–50
Humidifiers, 99
Hunger pangs, 23, 57–58, 118, 133
Hyaluronic acid, 185
Hydration, 82, 136, 181–82
  constipation and, 64–65, 82
Hydrolyzed collagen, 88
Hygiene hypothesis, 151
Hypnotherapy, 73
Hypothalamus, 23, 24

Ibuprofen (Advil, Motrin), 168
Ice cream, 54, 56, 115, 117
Immune system, 11–13, 26
  dirt and, 150–52, 159–60
Immune tolerance, 11–12
Immunity, 11–13
Immunoglobulin A (IgA), 149
Imodium (loperamide), 68
Indoor air quality, 166–67
Infants, 147–48, 158
Infertility, 146
Inflammaging, 14–15, 37, 40, 110
Inflammation, 9, 10, 12–13, 39, 116
  of the skin, 89, 90, 93, 97
Inflammatory bowel disease (IBD), 6, 79, 110–11, 125
In-Place Marches, 192
Insoluble fiber, 19, 40–41, 43, 73, 174
Insomnia, 130, 134, 136, 137. *See also* Sleep deficiency
Insulin-like growth factor 1 (IGF-1), 90
Integrated pest management (IPM), 166
Intestinal permeability. *See* Leaky gut
Inulin, 19, 43, 108, 150
Irritability, 13, 117, 138
Irritable bowel syndrome (IBS), 42, 71–74
  antidepressants, 25–26
  causes of, 72
  diet for, 42, 56–57
  managing, 72–74
  probiotics for, 18–19
  sleep and, 134
  stress reduction for, 122, 125
  symptoms of, 71–72

Jet lag, 136–37
Jobs, Steve, 32
Joints and microbiome, 109–11

Kefir, 19, 176
Keratin, 86
Kidney disease, 130

Kimchi, 19, 44, 176
  recipes, 215–17
Kitchen. *See* Diet
Knee arthritis, 108
Knitting, 123, 182
Kombucha, 176

Lactase, 54–55, 71
Lactase supplements (Lactaid), 71
Lactiplantibacillus plantarum LP-115, 17–18
Lactobacillus, 17–18, 44, 142, 146–47, 148, 154
Lactobacillus rhamnosus, 147
Lactose intolerance, 54–55, 68, 70–71
Leafy greens, 34, 92, 93, 175, 178
Leaky gut, 9–11, 28, 116–17, 169
Leg Stands, 106, 190
Leptin, 23, 133
Libido, 143–44
Liver disease, 50, 170
Loitering on the toilet, 76
Loneliness, 32–33, 129
Long COVID, 4, 16
Lower esophageal sphincter (LES), 55–56
L-theanine, 46
Lullabies, 140
Lunches, recipes, 202–4
Lycopene, 91–92, 206

Main dishes, recipes, 222–29
Makeup brushes and sponges, 99
Margarine, 49
Marine Stewardship Council (MSC), 218
Maroon-colored stool, 63
Marriage
  dementia and, 32–33
  leaky gut and fighting couples, 116–17
Maskne, 97
Mattresses, 140–41, 144
Meditation, 34, 120–24, 126
Mediterranean diet, 36–37, 39–40, 111
Melatonin, 132–33, 139
Melatonin supplements, 133, 136–37
Menopause, 1, 109, 142
Mental health, 113–27. *See also* Stress
  food and mood, 117–18
  leaky gut and fighting couples, 116–17
  microbes and mood, 114–16
Metabolites, 20, 23, 26–27, 115
Microbiome, 2–3, 6
  autism and, 153–54
  bacteria and, 6, 7–8
  bones and, 109
  children and, 151–53, 156–57
  cleaning products and, 161–62
  depression and, 27
  diarrhea and, 69
  diet and, 38–40
  dysbiosis, 8–9
  exercise and, 102–3
  inflammaging and, 14–15
  joints and, 109–11
  microbes and mood, 114–16
  migraine headaches and, 27–28
  muscles and, 108
  overview of, 5–6
  serotonin and, 24–25
  sex and, 143–44
  skin and, 85–86, 88–91
  sleep deficiency and, 134–35
  stress and, 115–16
Microflora hypothesis, 151
Microglia, 131
Migraine headaches, 27–28
MIND (Mediterranean-DASH Intervention for Neurodegenerative Delay) diet, 37–38
Mindful eating, 57–58, 60
Mindfulness, 57, 73, 121–22, 182
Miso, 19, 44, 176
Moisturizers, 87, 93, 96
Monounsaturated fats, 42–43, 48
Monterey Bay Aquarium Seafood Watch Program, 218
MOOCs (massive open online courses), 34

Mood, 6, 21, 31
  exercise and, 102
  food and, 117–18
  microbes and, 114–16
  serotonin and, 24–25, 115, 118, 143
Motivational stickers, 126
Mountain Climbers, 189
MSG (monosodium glutamate), 28
Multiple sclerosis, 29
Muscle mass, 108
Muscles and microbiome, 108
Mushrooms, 50
  recipes, 202, 204, 226–27
Music
  for exercising, 112
  for sleeping, 140, 144

Napping (naps), 138
Naproxen (Aleve), 168
National Heart, Lung, and Blood Institute (NHLBI), 37
National Institute on Aging, 37–38
National Institute on Alcohol Abuse and Alcoholism, 169
National Institutes of Health's Human Microbiome Project, 6
Nausea, 8, 10, 77, 114
Necrotizing enterocolitis, 148–49
Negative self-talk, 124
Nervous system, 22, 114
Neurodegenerative disease, 28–30
Neurotransmitters, 23–25, 115, 122
Night shift workers, 141, 184
Night sweats, 142
Nitrates, 47–48
Nitrites, 47–48
Noise level in bedroom, 142
Nonprescription drugs, 168–69
Nonsteroidal anti-inflammatory drugs (NSAIDs), 168–69
Nontime, 32
Nursery. *See* Children
Nutrition. *See* Diet
Nutritional psychiatry, 116
Nuts, 175, 178
NYU Langone Medical Center, 1–2

Oatmeal, 40, 42, 52, 56, 73, 175, 177
  recipe, 200
Obesity, 6, 13, 23
  children and, 150, 161
  role of diet, 49, 52–53
  sedentary lifestyle, 31, 104
  sleep deficiency and, 130–31
  weight loss, 41, 52–53
Oligofructose, 43
Olive oil, 48, 178
Omega-3 fatty acids, 31, 48, 93, 179, 185, 218
Online exercise classes and videos, 105
Oral cancer, 171
Organic produce, 167–68, 172
Osteoarthritis, 14, 109–11
Osteoporosis, 6, 109
Overhead Presses, 194
Over-the-counter drugs (OTCs), 168–69
Oxytocin, 119–20

Paints, 166–67
Pancreas disease, 63
"Pandemic pounds," 4, 52, 121–22
Parkinson's disease, 29, 46
Peppermint oil, 74
Performance anxiety, 115
Perimenopause, 142
Perinatal depression, 147
Pesticides, 165, 166, 167, 168
Pets and children, 152–53
Physical activity. *See* Exercise
Phytonutrients, 44–45, 177–78
Pickles, 19, 44, 176
Pillowcases, 144
Pink bismuth (Pepto-Bismol), 62, 68
Pituitary gland, 24
Plasticizers, 165
Polycystic ovarian syndrome (PCOS), 146
Polyphenols, 43, 47, 50, 93, 170, 177

Polyunsaturated fats, 48
Positive self-talk, 124
Postbiotics, 20
Postpartum depression, 147
Potty training, 81–82
Prayer, 120–21
Prebiotics, 19, 43–44, 59, 153
  best sources of, 19, 43, 176–77
Pregnancy, 64, 93, 146–49, 152
  hemorrhoids and, 75–76
Preservatives, 8, 10, 47–48
Preterm infants, 148–49
Probiotics, 17–19, 43–44, 59
  best sources of, 19, 44, 176
  for bone health, 109
  for children, 152, 154–56, 158
  for dogs, 153
  for hot flashes and night sweats, 142
  for IBS, 74
  for long COVID, 16
  for migraines, 28
  during pregnancy, 147, 152
  for skin, 86, 91, 94, 95, 97–98
Processed foods, 8, 10, 31, 35, 38, 51, 56, 90, 93, 95, 156–57, 174
Processed meats, 47–48
Protein, 49–50
  after workout, 60, 181–82
  for building muscle, 108
  recommended amount of, 49–50
Protein smoothie, 181–82
  recipe, 240
Psoriasis, 89, 91, 94
Psychotherapy, 124–25. *See also* Cognitive behavioral therapy

Quad Stretches, 192

Radiation therapy, 10
Radicals damage, 44–45
Rainbow colors of produce, 59, 177–78
Raisin meditation, 126
Rashes, 13, 86, 89
Reading, 31, 34
Recipes. *See* Gut Reno recipes
Rectum, 62
Red meat, 31, 39, 50, 79
Red stool, 62
Red wine, 46–47, 92, 170, 177
Relaxation, 119–24, 139–40
Resistance training, 105, 180–81, 190
Resveratrol, 91–92
Retinoids, 93
Reverse Crunches, 188–89
Rheumatoid arthritis, 10, 110
Roll Ups, 193
Rosacea, 89, 91, 94, 98
Russian Twists, 188

Saccharin (Sweet'n Low), 51
Saccharomyces boulardii, 17, 154
Salads, recipes, 205–8
Salmonella, 7, 66
Salt, 8, 38, 174
Sarcopenia, 6, 108
Satiety, 23, 57–58
Saturated fats, 48, 49, 50, 178
Sauerkraut, 19, 44, 176
Scissor Kicks, 193
Sebum (skin oil) glands, 87, 90, 91, 93, 95, 97
Sedentary lifestyle, 31, 103–4
Seeds, 175, 178
Selective serotonin reuptake inhibitors (SSRIs), 24
Self-affirmation, 124, 126, 182
Self-criticism, 124
Semivolatile organic compounds (SVOCs), 165–66
Serotonin, 24–25, 115, 118, 143
Sex, 143–44, 146
Shampoos, 94–95
Shinrin-yoku, 124
Short-chain fatty acids (SCFAs), 41, 90, 103, 122
Side (Lateral) Raises, 195
Side sleeping, 135–36, 144
Simethicone, 70

Single Leg Standing, 106, 190
SIRT1 (Sirtuin 1) gene, 92
Skin, 85–99
 aging and, 87–88
 diet for, 88, 91–93
 gut and, 94
 Gut Reno beauty regimen, 184–85
 Gut–skin axis, 85–87
 microbiome and, 85–86, 88–91
 probiotics for, 86, 91, 94, 95, 97–98
 structure of, 86–87
Skin cancer, 45
Skin care, 97–98, 99
Skin microbiome, 85–86, 91, 96–97
Skintense, 99
Skin-to-skin contact with infants, 147–48, 158
Sleep, 129–44
 alcohol and, 170
 bedroom environment, 140–43
 bedtime routine, 138–40
 benefits of, 129, 130–31
 for children, 157
 digestion and, 135–36
 Gut Reno sleep plan, 183–84
 jet lag, 136–37
 melatonin and, 132–33
 recommended amount of, 130
 for skin health, 99, 185
Sleep apnea, 130, 134, 136
Sleep deficiency, 8, 130–31, 136–37
 bacteria and, 133–35
Sleep–wake cycle, 131–33
Small intestinal bacterial overgrowth (SIBO), 74, 94, 170
Small intestine, 6, 9–10, 23, 26–27, 40–41, 74
Smartphones, digital detox, 120, 126, 184
Smoking, 29, 104, 170–71
Smoothies, recipes, 240–41
Snack bars, 59, 60, 181–82
Snacks (snacking), 52, 59, 60
 recipes, 234–39
Social isolation, 32–33, 129
Social relationships, 32–33, 183
Socks in bed, 141
Soda, 51, 70
Soluble fiber, 19, 40–41, 73, 174
Sorbitol, 66
Special diets, 53–57
Sperm, 146
Spices, 46, 179
Spicy foods, 66, 68
Squats, 190
Stage fright, 115
Standard American Diet (SAD), 8, 31, 35, 38, 39–40, 51, 156
Standing desks, 107
Stomach butterflies, 24
Stool, 61–62
 color of, 62–63, 82
Stool softener, 65
Stool testing, 80
Straight-Arm Planks, 189
Straight Leg Raises, 193
Stress
 constipation and, 64
 cortisol and, 24, 114, 115, 118
 exercise for, 102–3
 food and mood, 117–18
 microbes and mood, 114–16
 microbiome and, 115–16
 overview of, 114
 skin and, 99, 185
Stress management, 119–24
 meditation, 120–24
Stretching exercises, 106
Stroke, 36, 48, 130
Sugar, 8, 35, 38, 50–51, 88, 90, 117, 156, 174
Sugar alcohols, 66, 70
Sunlight, 45, 91–92, 98, 184–85
Sunrise alarm clocks, 143
Sunscreen, 98, 184–85
Supportive touch, 119–20
Suprachiasmatic nucleus (SCN), 132–33

Sweat glands, 87
Synbiotics, 19–20

Teeth brushing, 59
Tempeh, 44, 176
Therapy, 124–25. *See also* Cognitive behavioral therapy
Tightrope Walk, 192
TMAO (trimethylamine N-oxide), 39
Topical probiotic extracts, 97–98
Trans fats, 48–49
Traveler's diarrhea, 67–68
Triceps Extensions, 194
Triclocarban, 160
Triclosan, 160
Tryptophan, 25, 117
TULA Skincare, 97–98
Turkey, 25, 225, 233
Turmeric, 46, 97, 179, 210
TV watching, 104, 120, 139
Type 2 diabetes, 6, 36, 46, 50, 130–31
Tyrosine, 118

Ulcerative colitis, 152
Ultraviolet light (UV), 45, 91–92
Unstructured time, 32
Upright Rows, 194
USDA (United States Department of Agriculture), 167

Vaginal microbiome, 146–47
Vagus nerve, 22, 23, 115, 120
Vegetable oils, 48, 178
Vegetables, 36, 37, 43
  fiber in, 175
  frozen, 47
  organic, 167–68, 172
  phytonutrients in, 177–78
  rainbow of colors, 59, 177–78
  recipes, side dishes, 209–17
  skin health and, 91–93
Vegetarian diet, 36–37
Viruses, 7, 66, 153, 155
Vitamin A, 92–93, 226
Vitamin B6 (pyridoxine), 118
Vitamin C, 45, 88, 91, 92
Vitamin E, 43, 90, 92
Vitamin K, 43, 226
Volunteering, 34
Vomiting, 77, 114

Walking, 105, 123–24, 180, 194
Weight loss, 41, 52–53
Weight-training, 105, 180–81, 190
White noise, 142, 144, 183
Whole grains, 36, 118, 174, 175, 178
Window treatments, 141
Work from home (WFH), 107
Worry dolls, 140
Wrinkles, 13, 85–88, 185

Yellow stool, 63
Yoga, 73, 106, 191
Yogurt, 19, 44, 54, 56, 59, 157, 176
  recipes, 198, 200, 232

Zen Corner, 113. *See also* Meditation

ABOUT DR. RAJ

Roshini Raj, MD, is a graduate of Harvard College and NYU School of Medicine. She is a board-certified gastroenterologist with an active practice and holds a faculty position as associate professor of medicine at the NYU Grossman School of Medicine. She has a successful second career as a television personality and medical journalist. In 2015, Dr. Raj founded TULA Skincare, a probiotic-based skincare line sold through Ulta Beauty, Neiman Marcus, Nordstrom, Tula.com, Amazon, and QVC.

Dr. Raj appears regularly as a contributor discussing a range of health topics on television shows, including *The Dr. Oz Show* and *The Rachael Ray Show*, and on networks such as CNN and Fox News. Dr. Raj has also appeared on many other programs and channels, including *The Talk*, *Today*, *The View*, *Live with Kelly and Michael*, *AC360*, *CNN Tonight*, *The Doctors*, *Wendy Williams*, *Steve Harvey*, and NBC News. She is the contributing medical editor of *Health* magazine and is quoted frequently in national publications, including *Cosmopolitan*, *Glamour*, the *New York Times*, and the *Wall Street Journal*.

Dr. Raj is the author of *What the Yuck? The Freaky and Fabulous Truth About Your Body* (Oxmoor House, 2010). She lives in Manhattan with her two sons.